# DECISION-MAKING IN HIGH RISK ORGANIZATIONS UNDER STRESS CONDITIONS

# DECISION-MAKING IN HIGH RISK ORGANIZATIONS UNDER STRESS CONDITIONS

ANTHONY J. SPURGIN AND DAVID W. STUPPLES

## CRC Press
Taylor & Francis Group
Boca Raton  London  New York

CRC Press is an imprint of the
Taylor & Francis Group, an **informa** business

CRC Press
Taylor & Francis Group
6000 Broken Sound Parkway NW, Suite 300
Boca Raton, FL 33487-2742

Printed on acid-free paper
Version Date: 20160317

International Standard Book Number-13: 978-1-4987-2122-6 (Hardback)

### Library of Congress Cataloging-in-Publication Data

Names: Spurgin, Anthony J., author. | Stupples, David W., author.
Title: Decision-making in high risk organizations under stress conditions /
Anthony J. Spurgin and David W. Stupples.
Description: 1 Edition. | Boca Raton : Taylor & Francis Group, 2016. |
Includes bibliographical references and index.
Identifiers: LCCN 2016000967 | ISBN 9781498721226 (alk. paper)
Subjects: LCSH: Decision making. | Job stress. | Management.
Classification: LCC HD30.23 .S7217 2016 | DDC 658.4/03--dc23
LC record available at http://lccn.loc.gov/2016000967

**Visit the Taylor & Francis Web site at**
**http://www.taylorandfrancis.com**

**and the CRC Press Web site at**
**http://www.crcpress.com**

Printed and bound in the United States of America by Publishers Graphics, LLC on sustainably sourced paper.

## No Man Is an Island

No man is an island entire of itself; every man
is a piece of the continent, a part of the main;
if a clod be washed away by the sea, Europe
is the less, as well as if a promontory were, as
well as any manner of thy friends or of thine
own were; any man's death diminishes me,
because I am involved in mankind.
And therefore never send to know for whom
the bell tolls; it tolls for thee.

**MEDITATION XVII**
*Devotions upon Emergent Occasions*
*John Donne 1573–1631*

# Contents

# Preface

We have been interested in the occurrence of accidents and the relationships of organizational operations to the accidents. Potentially a lot more can be done to reduce the number and severity of accidents. In this, we have been confirmed in the study of ranges of accidents that have occurred in different industries and countries. Our studies in this field are reflected in this book.

The world is strongly affected by accidents—from local incidents to more global accidents that lead to the deaths of thousands of people and economic stresses in impacted countries The accidents range from a single death because of a traffic accident to a whole country affected by the arrival of a tsunami that destroys houses, kills hundreds of people, and floods the countryside with salt water that restricts the capability of the land to produce food for the population, possibly leading to hunger and malnourishment for years.

History tells us that every country is affected by accidents, and it appears that we cannot escape from being burdened by these events occurring and affecting us, individually, with pain and financial hardship.

We have been involved in the study of organizations and the impact of accidents on organizations. The idea of an accident being a random event over which we have no control and are in no way capable of preventing is an undesirable one. Closer examination of accidents reveals that if different decisions were made, better studies carried out, and the attitude of key persons were open to think more clearly about the likely consequence of actions, then the probability of an accident occurring could be heavily reduced.

The managers of more complex systems are required to be better prepared than for simple systems. From our background in the study of systems and related accidents, we have become more aware that the frequency of accidents in all kinds of situations can be reduced. It may not be totally possible to eliminate all accidents, because one lacks the ability to control everyone and everything. However, one can develop methods and techniques to enable persons directing and carrying out actions to minimize the potential for inadvertent decisions and actions being taken that lead to accidents.

This philosophy flies in the face of the "normal" approach to accidents (Perrow, 1999). Perrow's model seems very much influenced by the Three Mile Island accident sequence. His point was that as people make mistakes, big accidents have small beginnings and failures are those of organizations. The main idea is that failures are built into society's complex and tightly coupled systems, and as such accidents are unavoidable and cannot be designed around. We believe that organizations can be helped to reduce the probability of accidents and we lay out the processes here to accomplish this.

Our reanalysis constructs a different view of an accident (see Chapter 7). Our studies of a variety of accidents indicate that there are key events in most accident sequences where if the key decisions are approached differently, it would be very likely that the accident would not occur. In the field of accident analyses, there are parallel situations in which one decision path leads to failure and another path leads to success. This shows that it is not inevitable that a single failure will lead to an accident. The awareness of the decision-maker of an undesirable situation can be increased by being subject to better background training. This training should make a difference and lead to a reduced probability of a situation escalating to a damaging accident. It is better to avoid an accident than try to recover from the effects of an accident progression. The progression of an accident can tend to obscure the root causes of the accident and make it much more difficult for the organization to recover.

Some accidents occur because people, carrying out simple tasks, fail to realize that the repeated operations cannot be continued forever without a change occurring (linear extrapolations have limits). A case that comes to mind is the South Wales Aberfan coal tip accident in which 116 children and 28 adults died at a school on October 21, 1966. Here the local coal mines continuously tipped waste from the coal diggings onto a pile that grew ever closer to the school. No one gave thought to the fact the pile of waste could slide. However, the coincidence of the buildup and rain ended with a slide/avalanche that covered the school and led to the deaths! Thought about the configuration should have caused someone to be concerned about what might happen and its consequence. The failure of someone considering the characteristics of the pile was one of the accidents, along with others, that drove us to try to improve the decision-making process. The inquiry (October 1966 onward) found that the coal board was at fault: "The coal board was at fault due to ignorance, ineptitude, and a failure of communication." This finding indicates to us that it is possible to move away from the coal board–type of organization to one that is more responsive to potential accidents!

Another example of simple failures to think about situations and consequences is the emergency evacuation of very ill patients from hospitals in the Boston area following the *Sandy* storm surge of October 2012. The evacuation was needed due to the poor location of standby emergency power in the basements of the hospitals without adequate protection from floodwaters. Clearly, the standby power source should have been protected from the floodwater by either placing the power source higher in the building or having adequate flood damage control. The press expresses great concern that nuclear power plants in Japan are not protected from tsunamis, but we forget to take simple actions in our cities to save people here!

Clearly, we need to think about the subject of weather protection as whole, including flooding. Tsunamis are known throughout the world and people need some protection from these events. Fairly regularly, large numbers of people lose their lives from tsunamis. A typical example is the Aceh tsunami of December 26, 2004, in which some 170,000 Indonesian people were killed or were missing (see Aceh Tsunami, 2004). This does not cover people who died in India and elsewhere, because of the Indian earthquake-induced tsunamis.

We are encouraged that the authorities in England decided to protect London and surrounding areas from the combination of North Sea storm surges coupled with high tide with the Thames Barrier (see Thames Barrier, 1982). It was estimated that the design basis of the barrier should exceed the highest floodwater height that could occur once in a thousand years. It is interesting that the English authorities decided to act to build such a barrier. The feature of the barrier was that it could allow ships to pass normally and could be raised in response to predicted weather conditions, like those imagined occurring once in a thousand years. Raising the barrier has occurred several times since it was built to combat significant but smaller water levels.

The probability of the Fukushima tsunami was about the same number, and Tokyo Electric Power Company, Incorporated (TEPCO) and the Japanese government decided not to act! The failure to act led to 20,000 deaths and 140,000 houses damaged/destroyed. Seems that it was not only a failure by TEPCO, but also of the Japanese and local governments to act. The Fukushima nuclear plant's losses affected TEPCO's viability with the loss of power generation and the future costs to protect the public from possible releases from the damaged fuel, including plant clean-up. This accident gives all management concerns about the cost of decisions that can lead to destruction of the company. It should be pointed out that further to the north, at a place called Onagawa, both the surrounding people and the nuclear plant were protected by the actions of the power company vice president, Yanosuke Hirai.

These examples have brought us to a point where we want to help reduce the probability of these kinds of events. We believe that this can be achieved by approaching how management decision-making can be improved. Clearly, there seem to be some shortcomings in what is being done now. There does appear to be a defeatist attitude in thinking that accidents cannot be prevented, whereas one can appreciate that people with well-prepared and -trained minds can take actions to help mitigate the effects of an accident or even prevent its occurrence.

We see that to accomplish this objective, several techniques need to be brought together. Candidates for top management positions need to be educated in appropriate technologies. Their training needs to be one of continuing exposure to testing their decision-making skills. The training process that is proposed is based upon Adm. Rickover's principles, as are applied by the military, and applied to train naval submariners, marines, and others. In addition, two more tools are proposed to be included. Control systems should be changed to match variations in accidents by using an understanding of Ashby's law of requisite variety (more on this later). The other is training on the impact of human behavior on human limitations in responding to developing accidents. From his research a Danish researcher (Rasmussen) developed categories of behavior called skill-, rule-, and knowledge-based behavior. People respond differently to circumstances depending on their degree of preparedness to a situation, so if they are well practiced in the requirements of a given circumstance, they are said to be operating in a skilled-based mode and correspondingly their errors are likely to be small. The response of people to accidents therefore depends on their preparation to combat a given accident. This topic is discussed in later chapters of the book.

Early in our studies it was decided that a cybernetic model of organizations was needed so that one could develop a good understanding of how organizations react to accidents. The normal organizational chart does not supply this dynamic interrelationship. The normal organization chart is a linear arrangement of who answers to whom and who the chief executive officer (CEO) is or who is a control-room operator, but this is not dynamic. What is needed is a different model of an organization, so Beer's model (Beer, 1975) was selected, since it gave these insights.

Analysis of various accidents can provide two contributions to the enhancement of decision-making in high-risk organizations. By studying the analysis of accidents, it can be shown that the accidents that have occurred in different industries have many of the same sources and are not so different from one industry to another. The other contribution is to make the point to management that it is not above making a poor choice of action that can lead to accidents and end in the demise of the organization. Management needs to be concerned with not only making a profit, but also reducing the probability of thoughtless actions leading to an accident that badly affects the public and the workers. As can be seen in Chapter 7, accidents can lead to destroying the countryside, death of large numbers of individuals, the loss of jobs, and the loss of viability of the organization.

With the application of appropriate techniques and with training, management can be encouraged to pay more attention to the details and enhance both the safety and economics of organizations. Admiral Hyman G. Rickover showed that with the right approach, this could happen in the operation of the US nuclear Navy in terms of safety and effectiveness.

# Acknowledgments

Tony thanks his wife, Margaret Jill Spurgin, for her support during the time that he was working on this book. Without her support, it is unlikely that it would have seen the light of day.

David thanks his wife, Magda, his son, Oliver, his daughter, Claudia, and all his friends and colleagues who, over the years, have put up with him burying himself in the world of research.

Both authors thank Dr. Natalia Danilova for her insights into the application of Ashby's law of requisite variety to fields beyond the narrow interpretation, to the world of decision-making in high-risk organizations. Her thesis is "Integration of Search Theories and Evidential Analysis to Web-Wide Discovery of Information for Decision Support," and her PhD was granted by City University, London.

Thanks also to Dr. Saleh H. Al-Ghamdi, for allowing the authors to use some material from his PhD thesis on safety, controller reliability, and organizational structures in airspace control in Saudi Arabia commercial airspace. His PhD was granted by City University, London. His thesis is "Human Performance in Air Traffic Control Systems and Its Impact on Safety."

Many thanks to Cindy Carelli, acquisitions editor, for her help and advice.

Many thanks to Heidi Spurgin for the design of the cover.

The contributions that Admiral Hyman G. Rickover (1900–1986) made to the field of safety by his work on submarines based on nuclear reactor power are remembered. A key element in the safety of the nuclear submarines is the skill of the crews, based on his training and selection methods.

Thanks to Lt. Col. Dom D. Ford, for insights into US Marine Corps training methods.

# Authors

**Anthony J. Spurgin, PhD,** is an engineer whose interest is in the improvement of safety and economics of various industries by the enhancement of managerial decision-making. He is also involved in human reliability analytical techniques for estimating probabilities of human action under accident conditions. He was educated in England and earned a BSc in aeronautical engineering. Much later he earned a PhD in engineering and mathematics from City University, London. After attending the university, Dr. Spurgin was a graduate apprentice at an aircraft factory and afterward, as an engineer, performed aeroelastic calculations for a number of aircraft, including a supersonic plane.

Subsequently, he moved to the nuclear energy field and was engaged in design station control systems for gas-cooled reactors (Magnox and AGRs) and developed digital simulations of stations to help in the design process. For a period, he was with the UK Electric Generating Board (CEGB) and was involved in the study of the dynamics of gas-fired and coal plants and their control systems.

Dr. Spurgin moved to the United States to work further in this area and was the principal designer of the control and protections systems for 3 Loop PWRs and later for high-temperature gas-cooled reactors. He was involved in the design of various test rigs covering steam generator behavior and loss of flow (LOCA) experiments.

Later, he became involved in human reliability assessment. He has been involved in developing the methods and techniques to enhance the safety of nuclear power plants (NPPs). He has been supporting efforts in this direction in various countries and aided the efforts of IAEA in enhancing NPP safety. He has published the book *Human Reliability Assessment: Theory and Practice* (2009, Boca Raton, FL: CRC Press).

**David W. Stupples, PhD,** focuses on research and development of radar systems and electronic warfare. For a number of years, he undertook research in this area at the Royal Signals and Radar Establishment (RSRE) at Malvern in the United Kingdom, followed by surveillance and intelligence systems research for the UK government. He then spent three years developing surveillance systems and satellites for Hughes Aircraft Corporation in the United States. In his early career, Dr. Stupples was employed in radar and electronic warfare by the Royal Air Force. Later, he was a senior partner with PA Consulting Group, where he was responsible for the company's consultancy work in surveillance technology and in risk analysis of safety-critical infrastructure.

# 1 Introduction

## 1.1 PURPOSE OF THE BOOK

The purpose of the book is to discuss the topic of decision-making as it applies to high-risk industries and how one can apply the lessons from accident analysis. This work is an attempt to improve the whole attitude of the business community as to what it really takes to select and train decision-makers in the control of these types of industries. It should be noted that the approach taken here can have applications to other fields of management endeavor. Even the word *accident* has the context of something unavoidable and uncontrollable and that despite actions taken by people, the inevitable will occur and we have no way to prevent it. This, of course, is not correct. Random events may occur; our position is that one should be in a position to minimize the effect of the incident. Here the words of the Boy Scouts are appropriate: "Be prepared." If an organization is prepared, then the consequences of an event can be reduced.

The authors have been employed in various industries and involved in studies of how decisions made during designing, locating, building, and operating nuclear power plant (NPP) units can affect subsequent operations. All of the decisions related to these actions can have an impact on the safety and economics of high-risk industries. High-risk industries are those industries where accidents can have a significant financial impact on the organization and affect the safety and health of the nearby population.

The authors have extensive experience in the fields of safety, design, and economics of plant operations of high-risk industries. The purpose of the book is to draw upon that experience, as well as insights derived from the study of a series of accidents in different industries, from nuclear to railways. A key element in our studies of accidents was the evaluation of the responses of organizations, both at the manager and operator levels—before, during, and after an accident. It was these observations that helped us develop our approach to the integration of various techniques, such as Rasmussen's skill, rule, and knowledge–based behavior (Rasmussen, 1997) and Ashby's law of requisite variety (Ashby, 1956) along with Beer's work (Beer, 1981) on cybernetic modeling of organizations.

From our studies, it became obvious that the top managers have been very much caught up in the economics of day-to-day running of the organizations and often have been trained to deal with these kinds of problems, such as MAs in the art of management. This approach was very much the standard for managing ordinary operations—involving capital cost and expenditures, profit and loss, and direction of staff to help ensure their productivity. However, it seemed that the bigger potential losses resulting from random accidents are not seriously embedded in the training of management personnel. It could be that many managers did not really understand the implication of probability estimates; a one-in-one-thousand estimate does

1

not mean that the event could not occur tomorrow! Many of the decisions taken by managers seem to ignore potential risk implications on the basis of their "feel" about the situation, rather than a realistic review of the situation, or the acceptance of advice from persons with better expertise/knowledge. A typical example of this is the *Challenger* accident (Chapter 7).

It appears that sometimes the high economic consequences of low-probability accidents on the cost of the operation, or even the survival of the organization, are not dominant considerations in evaluations made by management. Often, in the evaluation of the risk of an incident the probability of such an event was estimated to be very low—less than 1 such event in 1000 years—and therefore could be ignored. This can lead to poor decisions. As one knows, even if the probability is low, the event may occur the next day. For example, the recent Fukushima tsunami event was considered to be a very low-probability event, but the consequences of failing to consider it possible was high (for both TEPCO, the government, and the Japanese people). The result was the loss of four reactors, destruction of 140,000 houses, and the death of some 20,000 persons. In addition, the land flooded by the tsunami was affected by saltwater contamination (which affects productivity of the land). A detailed discussion of the Fukushima tsunami-induced accident is in Chapter 7.

Not all of the accidents are due to the coincidence of a severe accident coupled with the failure of staff and management to have taken the necessary actions. Sometimes, the situations are mundane, but the situation is not improved by the failure of management in planning for such situations by instructing and training their staff in the appropriate actions to be taken and rehearsing their actions. In the case of the Fukushima accident, the management was aware of the possibility of a tsunami larger than the protection offered by the then current seawall, but decided not to act to increase the height of the wall (McCurry, 2015). However, no one has held the Japanese government accountable for its failure to protect the people affected by the tsunami; see earlier text. It is possible to protect against floods (tsunami induced) by the use of seawalls and barrage schemes (Thames Barrier, 1982) designed to combat the coincidence of high tides and storm surges. The seawall at Onagawa NPP protected both the nuclear plant and the local population (Maeda, 2011)!

In the study of accidents, we found that a key element in the failure to control or mitigate accidents was in the training, education, and experience of management, along with a better understanding of the processes used in their specific industries. There also seemed a failure to support their decision-making with advice from competent individuals. Examination of many accidents revealed cultural problems that prevented advice being accepted. These cultural problems are not limited to certain societies. These effects turn up in a number of accidents, such as the Fukushima accident and also in the *Challenger* accident. The cultures may be different, but the results turn out to be similar.

The analyses of accidents can be very enlightening, but to study the decision-making process of management one needs to consider the whole process of decision-making, starting with an understanding of the human control system (management) governing how an organization operates and takes actions through information and instructions passed from management through senior staff to operators.

We have used Beer's *cybernetic model* based on the human body (Beer, 1979) to help us understand how management relates to operations. Beer's cybernetic model provides a structure to enable one to relate organizational aspects of the roles of managers to control-room operators, information generation and advice for managers, decision-making and planning, training and maintenance of equipment, and so on.

In our studies, we have developed Beer's model to show the relationship of how accidents affect the response of the management function as an input along with the role of the regulator of the industry, such as the Nuclear Regulatory Commission (NRC). It should be pointed out that not all regulators act in the same way; the imposition of rules, regulations, and controls may not be the same, but they all have an impact.

## 1.2 POTENTIAL USE OF THE BOOK

The primary objective of the book is to discuss issues associated with decision-making in high-risk organizations and introduce tools that could be used to help in the decision-making process. The hope is that this can contribute to the reduction in the number and severity of accidents. It also discusses how managers can improve the decision process to reduce the probability of an accident causing significant harm. The book covers a number of techniques that need to be integrated into the management processes to enhance the capability of organizations to deal effectively with accidents. Accidents may not be single events that the management can easily deal with, but can often be complex events with different stages and different interactive events, which make it difficult to deal with, even to the point of negating the best efforts of the plant personnel to terminate or mitigate the effects of an accident.

For example, the earthquake that affected the Fukushima NPPs was responded to appropriately and safely initially; however, the impact of the large tsunami caused by the earthquake caused effects that led to damage to the reactor core and the release of radioactivity, despite the best efforts of the station personnel.

This book could form an integral part in the training and education of managers and staff of high-risk organizations. The approach taken here is that training for all management personnel starts during the early stages of potential management personnel selection and continues to be more comprehensive during their progression within the organization. This process mirrors the training programs of the military, where continuous improvement is the lifeblood of success in dealing with rapidly changing environments as opponents react to one's tactics.

This book could be the basis of a set of lectures given at business schools. The business schools' curriculum focuses on a number of issues, such as accounting, organization, and so on, but the impact of technology is not just in the area of tools for monitoring the costs, inventories, and so on. Technology can bring with it an increased risk because of the improved efficiency of the processes. The resulting reduced thermal capacities of equipment can lead to reduced response times to disturbances; this emphasizes a need to improve the awareness of management of the requirement for both their own training and of that of their staff.

A central role in the safety of high-risk organizational (HRO) operations is the role of regulators to monitor HRO operations to ensure that these organizations follow the rules set up by the regulators and fulfill their responsibility to the general public.

## 1.3  PHILOSOPHY OF THE AUTHORS

It is our philosophy that people do make mistakes, but the resulting effects can be reduced or mitigated by better decision-making on behalf of management in coping with accidents. However, it does require effort on behalf of all persons involved in an operation. The responsibility to be aware of what can be done and what should be done rests on those in management positions, since they make the key decisions. It is they who have the power to make decisions. Subordinates may be in a position to observe what might be done to ameliorate the situation, but the managers are empowered to make decisions; this is a fact of life. One can see from accident analyses many situations in which the advice of others is not taken. The purpose of discussing accidents in a wide range of industries is to illustrate this point.

One comes across opinions held by persons that accidents cannot occur in our industry—we are too careful, and our operators are too well trained for this to happen. Subsequently, a major accident occurs and blame is then shifted to the failure of the operators to take the correct action and that they needed more training (Three Mile Island [TMI]), or a key piece of equipment failed (Gulf of Mexico oil release). Sometimes, a large accident is avoided by building a seawall, only to receive jeers from people on the stupidity of doing so (Onagawa), versus the failure of clever persons not to take this action, and then being party to a severe nuclear accident (Fukushima, TEPCO).

We ask ourselves, How do we, as a group, improve what we are doing to help to minimize the effects of accidents? Clearly, technology has improved in terms of detection and taking action. Information on the state of the plant has also improved. The training in the capability of operators to deal with emergencies has improved in the nuclear industry, in part due to the TMI #2 accident, with improved training and the use of plant simulators. It is not clear that other industries have learned from this experience.

The whole training approach toward the NPP operators has improved and continues to do so. However, the application of these techniques—the use of simulators, development of procedures, information presentation, and continuous training in the science of nuclear power—does not seem to be applied to those who might be more in need of it, the decision-makers. Exposure of the decision-makers to the details of past accidents and their consequences in a variety of industries is most necessary to influence their attitudes toward being prepared for accidents, especially those different in characteristic than ones that their industry has already addressed.

Bit by bit the nuclear industry has grown in its effective response toward both safety of the public and economics of operation! This does not mean that the core of the industry should rest on past achievements and experience. Too quickly, memories and knowledge disappear; it is good to remember the statement, "The price of liberty is ever-present vigilance." The same principle holds for safety and economics.

Our approach to improving the decision-making ability of managers is based upon the idea that better preparation of managers can decrease the probability of accidents occurring and enhance recovery from the effects of an accident. Improving the performance of managers can be achieved by a better training process—making managers more aware of risk exposure, what can be done about reducing risks, and increasing awareness of the dynamics of organizations and how that determines consideration of pre-, mid-, and post-accident activities.

In the case of preparation of persons for management posts, one can look at the training of military personnel (Navy and Marines). One can see the value of use of simulators in the preparation of operators. In the case of management, the emphasis should be on the improvement in decision-making by exposure to increasingly difficult scenarios. One can see that military training uses simulations of battlefield conditions, from small-scale to large-scale operations. Here, as in the emergency plant conditions, the objective is to prepare the individual to make decisions and take actions. The test scenarios are gradually increased in difficulty and complexity, so that the capability of the individual is enhanced. Furthermore, one sees which individual is more capable of dealing with these situations; hence selection of a manager can be based on objective measures of performance.

Tools and processes are needed to improve the whole business of managing high-risk operations. We have identified a number of these tools and methods. Clearly, training in the technology of the operation is one, and as mentioned, simulations of various accident scenarios and understanding of organizational dynamics—how they operate, and role of the support staff in terms of actions to be taken and advice given. In addition, one needs to understand the behavioral aspects of persons when exposed to an accident. It is also extremely necessary for decision-makers to understand that the dynamics of processes can evolve during an accident, and to understand how to better control the accident progression. To be effective, all of these methods and techniques need to be understood, integrated, and applied.

The following methods and the aforementioned processes are covered in detail in the following chapters:

Chapter 3: Cybernetic Organizational Model: Beer's Viable Systems Model
Chapter 4: Ashby's Law of Requisite Variety and Applications
Chapter 5: Probabilistic Risk Assessment
Chapter 6: Rasmussen's Human Behavioral Groups
Chapter 7: Case Studies of Accidents for Different Industries
Chapter 8: Lessons Learned from Series of Accidents
Chapter 13: Investment in Safety

The basis of the approach taken by the authors to improve decision-making in high-risk industries under stress is the integration of the aforementioned tools and methods, plus the effect of the lessons learned from the accident case studies. The case studies show that accidents are not exclusive to a given industry or organization. They show that there is a need for management to approach the issue of accident management in a more organized and trained manner.

# 2 Background

## 2.1 INTRODUCTION

This book deals with organizations, management, accidents, and impact of technology on the safety and economics of operating high-risk plants. It is also appropriate to examine what has happened in the last number of years to give some background to the effect of technological changes as well as the impact of management decision-making on organizations. We tend to forget the technological changes that have occurred and their influence on the state of affairs.

Technology has advanced over the past years starting just before World War II (WWII). The advances in technology have been tremendous and have had an effect on governments, the public, and on industries. One thing is how advances have been seen and acted upon by management to try to produce goods and services more efficiently. Governments, such as for defense, and the corresponding industries have also participated in the use of advances of technology (e.g., rockets). One specific technological advance is the use of space satellites for navigation and monitoring of activities on the ground (troop movements, etc.); this technology has rolled over to the general public to find locations.

## 2.2 APPLICATIONS OF TECHNOLOGY

There have been a number advances in the application of technology. It is not our intent to describe all possible applications and the details of such, but remind the reader of some of the advances and their applications. As one can appreciate, not all advances are without disadvantages, but this is usual in the world of humans.

### 2.2.1 SUBMARINES

This is one interesting technological change that illustrates the impact of these technological changes that occurred in the period since the end of WWII. This change has had a profound effect on the world and occurred in the world of submarines. This is just one of the changes; others have had similar effects, if not quite so influential.

Submarines have been around since the time of the American Civil War and were basically surface ships that submerged for a short period. The reason for their submerging was to be able to approach surface ships unobserved with the purpose of sinking them. A lot of technology went into trying to improve the performance of submarines, such as the use of better batteries and snorkels or breathing tubes, but these were just small incremental changes. On the other hand, adversaries developed sonic detectors, so that they could detect the submarines and destroy them, before they could cause damage to the ships.

The proposal to power the submarines with nuclear reactors caused a complete rethinking of the role of submarines. The first was the idea of allowing the true submarine to be developed. It was now not a modified surface ship and its hull shape reflects that. Also now it could remain submerged for months rather than hours! The next development was to modify its purpose to include launching ballistic missiles to attack land bases. The next was an attack version designed to defend these ballistic submarines (*boomers*) by attacking enemy attack submarines; but they could still perform the old tasks of sinking "battle" groups of aircraft carriers and other ships.

What is seen here is the first impact of the technology, the change in the characteristic of the submarine from a pseudo surface vehicle to a real submarine! However, there are other technological implications related to the new submarine. First, the submarine has to be safe, so that it can operate. Then it has to be reliable, both in equipment and importantly, humans running the submarine. The net result of both of these aspects is the development of the fields to produce these changes. The person identified as the leader of the whole field was Admiral Hyman G. Rickover. What a fantastic achievement: to not only introduce the nuclear reactor as the power supply for submarines, but to see what was needed to ensure that the submarines were operated successfully. He pushed industries to develop reliable equipment, but also he developed techniques.

The objective of the aforementioned commentary is to indicate that technology has changed over the years in the field of naval warfare, but this is not the only area. One needs to consider how technology has changed and what are its implications in the case of physical changes in the equipment but also how people and society approach these changes. Clearly these changes can affect how financial and safety situations are addressed. The methods of solving mathematical problems have also advanced from the use of slide rules to high-speed super computers. To make use of the "machine's" methods of solving mathematical problems, one needs to advance the human side of enterprises. One needs to train people to be able to make use of these capabilities and integrate them into operations.

The key element in any organization is management, and one needs to pay attention to their selection and training.

### 2.2.2 Aircraft Development

Almost everyone knows that airplanes have changed over the years. The changes are in both materials for airframes and in propulsion systems. The airframes have changed from wooden elements (spars, ribs, and fuselages covered with cotton sheets), then to aluminum structures performing the same functions, followed by carbon filaments embedded in polyurethane or equivalent materials. The airframes and wings are now the equivalent of modern buildings made of concrete reinforced with steel rebar. Of course, the weight is quite different, since aircraft performance is dominated by weight considerations!

In more modern times, the commercial success of an aircraft depends on a number of things, but central to the success are the ideas of the designers and how quickly they see the developments proceeding. The development of the jet engine was key to both military and civilian planes. It was seen early on that the jet engine was a step forward in speed, and it was adopted by the military. The civil development of jet engines was slower, with the propjet being adopted initially, which became popular. The limitation of the propjet was the propeller, because tip speeds had to be kept below supersonic speeds. The pure jet engine became dominant when it was realized that flights could be faster with pure jets and the manufacturers had improved fuel efficiency of the jets. The reason was the capability of making more flights per plane, a question of cost.

The size of the market is a factor. A manufacturer in a small country, such as SAAB, is at a disadvantage, even with the support of its government, despite the quality of its aircraft. This feature has led to the United States dominating the military aircraft field and being a strong force in the civilian field. In the civilian field, Boeing and Airbus are still in competition, but where are De Havilland, Hawker, Vickers, AVRO, and so on? Of course, Russia and China are different and their governments can control their markets. China has an arrangement with Boeing to exchange design data, while buying Boeings, but in the end will produce their own airplanes!

Not all technology has a smooth path. Rolls-Royce's fan jet engine tried to use an early form of plastic blades for the fan and it proved disastrous and they had to replace the fans with titanium blades, which were heavier. The better, lighter-weight blades were developed later! The decision was made by a less experienced project manager. Poor management decisions place organizations in tenuous states and this one could have led to the demise of Rolls-Royce's aircraft business.

### 2.2.3 COMPUTERS

Here the advances have been staggering and have been in the fields of computer manufacturing, software, and range of applications that have had their effect on every industry, organization, and life in general. It is almost impossible to cover the fields, they are so extensive: from changes in how business is carried out, controlled, and how goods are sent out—the handheld cell phone, the personal computer, the commercialization of personal interactions (Facebook, etc.).

The computer field has led to the development of simulators, simulations of industrial operations, the ability to analyze genomes to enable the sources of diseases to be understood, and so on. Industrial processes can be controlled and monitored more effectively. In addition, the costs of organizations can be evaluated and weaknesses in cost controls identified.

### 2.2.4 INSTRUMENTS

This is another area of development without which many of the advances in computer technology would be limited. Instruments are the sensors (the eyes and ears), which enable information about processes to be gathered. Applications of computers to analyze heart rhythms of patients cannot occur without pressure, temperature, and oxygen sensors. Computers need input values to operate upon to determine what to control, information to print, and so on. In the case of closed loop systems, what actions need to be taken by the actuators to achieve the requirements of the process/system are determined by signals generated by sensors and processed by computers. The accuracy and responsiveness of the sensors is critical for the process safety and its economics.

## 2.3   COMMENTS

The rapid development of technology can present difficulties to managers, especially those who have been trained in earlier periods. This is particularly true in the fields of electronics, computing applications, and software, where the pace of change is so rapid. Under these conditions, managers should probably return to school to receive additional training in these topics to help them make decisions in fields affected by these advancements.

Even older industries are being influenced by technological developments. Design processes are changing and components are being constructed in different countries and then assembled in the "home" country. Examples of these methods of working are Boeing and Airbus. These changes are very dependent on highly technical processes involving reliable data transfer, good communications, and software/mathematical tools. In addition, these methods call for deep coordination among designers, constructors, and various management teams via spatially connected interactive displays. These operations can only be carried out with the aid of extensive computer programming and testing. These operations present interesting problems for managers to follow, review, and control. If they are without the knowledge of how the tools work together and do not understand the designs, it seems impossible for correct and timely management decisions to be made. The old management method of being concerned with costs and manpower is now being augmented by concerns about technology, and the ability to recruit and keep technically competent staff, without which the organization will not succeed.

The adaption to technology has been a test for managers. It is equally a test for staff and their training. The world of manufacturing was, in earlier times, governed by simpler mechanical operations of drilling, turning, and even welding. Jobs like this are available, but many industrial operations now require a deep understanding of computer technology. One question: Is the educational system capable of training students for this new world?

So the world of education has to respond to this new world, and so the same concerns applied to managers now include educators. They need to modify the courses to include technological development; just to train students in programming is insufficient. By the time these students graduate, the world of computing has moved on and the methods have modified and changed, because change is occurring so quickly. The students need to understand how computers work and their basis. Concepts need to cover the fundamentals of computers like artificial intelligence and such tools. This is one estimate of progress; of course, one cannot completely define what progress means! None of the developments that have been made were entirely intuitive at the time.

The authors are both engaged in different aspects of technology and can appreciate the advancements in various industrial fields, and how technology has led to large changes in industry and in organizations. Attention is paid in this book to some of the advances and the implication of those advances on the decision-making capabilities of managers.

# 3 Cybernetic Organizational Model
## *Beer's Viable Systems Model (VSM)*

## 3.1 OVERVIEW

The purpose of this chapter is to present information on the viable systems model (VSM) developed by Beer (1985) and its application to better diagnose management systems. The hierarchical methods to depict management structures do not help one to understand how these operations actually function. VSM is a method to enable understanding of management dynamics of organizations, based upon cybernetics. The key word in VSM is "viable": capable of maintaining a separate existence. If one considers the roles of the various parts of an organization, one can quickly recognize that some parts make decisions, others plan operations, and yet others carry out those plans. Between these parts, there are communication channels transporting information about the processes being operated on and instructions to operations personnel to increase or decrease activities. Beer recognized these relationships as being similar to the detailed actions and responses of human and animal bodies; in other words, the same principles being used to understand how animals operate were relevant to the diagnosis of human organizations.

This chapter will cover control systems concepts so that one can appreciate how simple controllers work, alongside with ideas about feedback and feedforward signals used in the control of processes. Some further, concepts related to controls are introduced so that the jump to the complexity of cybernetics is more easily understood. Modern technology shows that more and more computers are being used in ways that resemble cybernetics. One example of this is the control of automobile engines. Automobile manufacturers have responded to the public needs by designing complex interconnected control schemes for cars to control pollution, prevent the release of noxious gases, and at the same time increasing fuel economy and power output. This has been done by developing controllers acting very much like cybernetic machines.

Cybernetics is the study of the structure of regulatory or control systems, which are seen in animals as well as in business systems. Cybernetics is closely related to control system theory. Cybernetics is equally applicable to physical and social systems. Here we are going to look at the application of VSM to air traffic control (ATC) in a foreign country, Saudi Arabia (Al-Ghamdi, 2010). VSM has been applied to a number of different organizations and accident situations by Spurgin in his PhD dissertation (2013b) as part of the analysis of accidents. He has applied it to a normal

US nuclear utility, and to the site organization trying to respond to the Tokyo Electric Power Company's Fukushima tsunami-induced accident.

VSM is built upon the ideas derived from cybernetics. VSM was proposed as a better way of understanding and diagnosing organizations to understand their behavior. The approach has been applied to manufacturing, food distribution (Walker, 1991), software development by Herring and Kaplan (2001), etc. VSM was applied by Beer to government operations in Chile under President Allende, circa 1970–1973, as seen in Beer (1981). This shows the diversity of VSM as a tool for diagnosing various management schemes and operations.

## 3.2   CONTROLLER DESIGN AND OPERATION

VSM in operation works in a very similar manner to an automatic controller. Information about the state of the organization's production is treated very much like signals flowing from a plant's transmitters. The controller receives the signals and sends out signals to the actuators to move to influence the system under control. Management receives signals from staff on the production levels and, following evaluation, sends commands to change the production level to balance the needs of the clients and the market.

The needs of the market are monitored over a period of time to see if production matches the needs of the market/public. If need be, the managers issue instructions to personnel to make further changes, as required. These changes are made until stability is reached, implying that production is matching the needs of the market. Of course, these can change in the market, raising or lowering the need of a particular product. Figure 3.1 shows a simple controller that can respond to set point changes or process system changes. This is like a company making changes to products or responding to public shopping changes.

The components of the control system are (a) a controller that embeds the control rules (or algorithm Fn [error]), (b) an actuator which affects the control action to produce a change in the process, (c) a sensor which detects changes in the process,

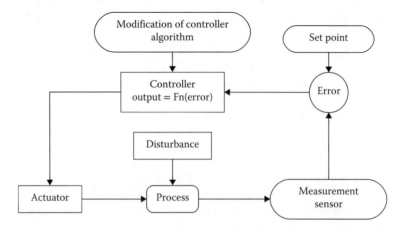

**FIGURE 3.1**   Simple one-loop controller and process.

and (d) an input set point to demand to establish the required state of the process. The output from the sensor is the feedback signal and is compared with the desired state setting. The difference between the two signals determines the error between the current state and the desired state. The controller acts through the controller rules to effect the necessary change in the process to reduce the error to zero and bring the actual state to conform to what is desired.

## 3.3   CONTROLLER RESPONSE

Figure 3.1 shows an approach to model controllers that incorporate algorithm adjustment. There are many ways that controller algorithms can be changed. One way is to do so automatically in response to changes in the process, like having different settings when the plant is at high power or low power. In the case of airplanes, the aerodynamic control of a plane would change settings according to altitude or Mach number. By this process, control would be more optimal and the plane more stable under varying conditions.

In the system shown in Figure 3.1, the actuator is likely connected to a valve. Valve movement leads to the process responding and the plant state moves toward the desired plant state as determined by the set point change. Also, if the plant is disturbed, the controller acts to return the plant state back to its desired state.

## 3.4   VSM SYSTEM

Following the description of how a normal controller works, VSM is now discussed. Figure 3.2 shows a simple version of a VSM model of a system, in which there is a central management body that determines policy and gives top-level guidance. Also there is a regulatory/control body that controls various activities at the working level (supervisors). Then there are the actual operational activities, from running a nuclear power plant (NPP) to shoemaking, tire production, etc. The environment represents the public, the physical environment, or even the government. Feedback occurs, and

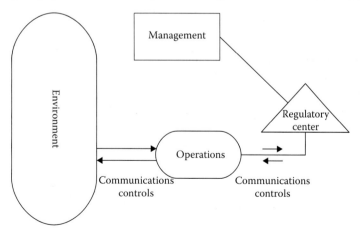

**FIGURE 3.2**   Basic VSM figure depicting key elements within approach.

actions can result to effect change in the activities of the plant/organization. For example, in the case of shoe manufacturing the public may change its taste from black shoes to red shoes and this would lead to a change in production rates of black and red shoes.

The regulator shown in Figure 3.2 operates in a similar manner to the controller shown in Figure 3.1. The set point could be related, say, to the number of shoes to be produced per month as set by top management of the shoe manufacturer. The regulator has a number of rules, which correspond to the algorithm of the controller and can be quite complex. These rules could cover such things as the color ranges of the shoes and the sizes and selection of materials. The rules may also determine the use of machines, targeted hours per shoe for manufacturing, and the length of run producing the shoes. The VSM model depicted here is a simplified model of a shoe manufacturing business. Feedback occurs from the operations function as to the construction and assembly of the shoes, as well as such things as the utility of certain machines to produce different kinds of shoes, the downtime requirements due to the need to maintain the machines, and the impact of shift changes of operating personnel. The VSM model structures can be expanded to include subunits with a similar structure to that shown in Figure 3.2. The expansion of VSM depends on the needs of the user.

A simple VSM model can be used to examine the relationships between the various key parts of the organization, that is, management, the control rules for operating the organization, the operations portion, and the environment (the public and other organizations can be affected by the organization's actions). The VSM models focus on both feedback and feedforward signals, which tie the various units together and make it possible for the whole to work. The dynamic aspect of the VSM model changes an organizational chart into an operating entity. An organization is unlikely to function successfully unless all of the roles played by all parts and their communications work together. Several organizations will be examined in Chapter 7 to show how failures within organizations can lead to accidents and even to the demise of organizations.

## 3.5   USE OF FEEDBACK IN VSM

Figure 3.2 can be viewed as a simple representation of a utility, but it can be considered to be a building block representing any organization. In essence, the management block represents the higher functions performed by the top management, such as optimization of the cost and safety effectiveness of the organization, reacting to information relayed by the rest of the organization, setting operational rules, and allocating resources. The regulator function covers the local management function (supervisors) that control the work product and reflect guidance from the top management. The operational block carries out the required tasks ascribed to the organization. The product of the organization then affects the *public* or the environment.

As mentioned earlier, feedback occurs at all levels of the organization, both forward and back. One item always of concern in any cybernetic situation is the quality and frequency of the signal (information or control). Each operating part of the organization needs to ensure the information neither overwhelms nor is deficient as far as the receiver is concerned. Some filtering is required; clearly management can make poor decisions if the quality of the information is defective. Equally, the top management has to make good decisions to minimize the risk to the organization or compromise the environment. An intrinsic requirement is that each part of an organization has to operate efficiently and safely, in order to meet the goal of operating in risky situations and still be economic.

In the VSM approach, the term *regulatory* is another word for controller and it is not to be confused with the term *regulator* as associated with the regulatory function of, organizations like the US Nuclear Regulatory Commission (NRC) or other similar bodies. In the government sense, the term *regulator* is used in the legal sense of a person or persons, who regulate(s) the actions of organizations to ensure they fall within legal constraints.

VSM operates in an environment, reflecting the outside world and its variability. The system senses and acts on environmental changes via the communications and control channels shown in Figure 3.2. This information is then fed back to the regulator and hence to the management function. The information and control channels are also used to characterize the health of the operations. The regulator examines this information and determines if it needs to be act on, or if the information is such that the management needs to be informed to change the operating rules. If the regulatory response is within its permit, then action is taken by sending messages to plant operations. However, if management is to be involved, information is sent to management and they in turn send instructions to modify the rules. Of course, the management might require more information before being in a position to change rules.

In manufacturing organizations, changes in regulation might be to increase production of, say, a certain type of shoe to match the demands of the environment (public). Information derived from the environment is analyzed to see if the planned quantity is likely to be required. If the design of the shoes is such that the market share of the organization is falling, then the action process is more complicated and management should be involved. They should analyze what steps should be taken, such as changing the design of the shoes in some way.

The VSM approach is a cybernetic approach to what is normally a hierarchical approach, where the center of operations is the management function, which implicitly includes the regulatory function. This means that decisions and regulation of the operation is held by management. In the VSM approach the management determines the rules, and the regulator controls the process via information obtained from the operations and the environment. The operations function carries out manufacturing, construction, and running the process. All features of the enterprise operate together to meet the needs of the environment (client). This concept is a much more shared system, with actions taken locally as appropriate.

## 3.6 COMPLEXITY OF OPERATIONS

The aim of an organization is to satisfy the needs of the public and to do so profitably. Of course, if one looks closer into what the public requirements cover, it soon becomes obvious that it covers more than just the impact of the delivery of a number of shoes. Although individual persons might be satisfied with the shoes produced by the organization, there are other organizations representing the populous call for constraints on the manufacturing of the shoes, so the environment depicted, in a simple way in Figure 3.2, implicitly covers the integration of all of these forces.

Thus the complexity of the system increases and the relationships between these outside entities have to be factored into the structure. However, the basic concept of an organic process is still maintained: other elements are introduced within the environment, the operations, regulation, and management. Information has to flow between the elements and actions indicated in the normal control manner. The importance of each part of the organization in satisfying the goals and objectives is still maintained. Failure or success in meeting any of the organization's needs or requirements can lead to failure or success of the enterprise.

## 3.7   ENHANCED VSM REPRESENTATION

VSM can be applied to higher or lower levels of representation of the systems; for example, if the company consists of a number of factories making different products, such as shoes, clothing, handbags, and so on, then all of these factories should be combined in order to consider the health of the overall organization. Equally, it can be used to represent the details of just the shoemaking operation. Each factory within a set of factories is shown in Figure 3.2, operating as a partial entity of the whole organization. In the case of multiple factories, there are as many VSM blocks as there are individual factories. All of these subunits come under the control/influence of a senior manager, but each individual unit would be self-regulated. In addition to the top management function, there would be other functions to ensure the balance between the subunits in terms of meeting overall management objectives for both the whole and the individual subunits.

Within the subunits, there could be subfunctions dealing with maintenance, finance, human resources, material purchases, etc. Each of these subunits would function as a self-regulated unit, but they cannot exist as an independent organization except within the main organization. For all of these organizations, units, and subunits to function, communications have to be timely, efficient, and accurate. The top management needs to receive credible information with which to make the best decisions. The speed of information has to be high enough, and filtering of the information is such that it is clear, accurate, and unambiguous. Equally, all persons within the organization should be informed about decisions affecting them and how they are expected to perform.

Most organizations operate as closed loops; however, they have react to how the markets change and be prepared to act to modify their products, either in quality or quantity. They are also required to respond to availability of raw or processed goods and also to monetary shifts. As in human body, the level of control depends on changes in the environment, so some organs of the body are under autonomic control and for others the higher levels of controls have to take over. In the absence of good and effective communications, the top management is required to step in and ensure the operations are carried out efficiently, but the organization is much more effective if each component operates efficiently without outside controls.

This consideration led to an enhanced VSM model to cover these aspects and is shown in Figure 3.3. It is appreciated that there is a need to produce some measure

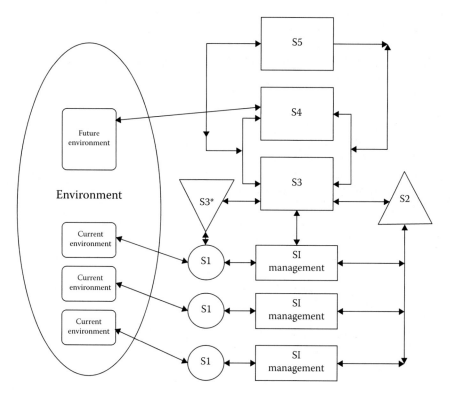

**FIGURE 3.3**  More complex version of VSM.

of stability, management-wise, between production units. Therefore, there is a need to link organizations to achieve this. Outside of the main part of the organization, there was a need for a regulator to cause each operation to act more in concert. Clearly, some degree of acceptance by managers and personnel has to accompany these control measures to make them effective. The stabilizing organization has to communicate with management and if necessary get it to understand and comply with the strictures. Figure 3.3 diagrammatically depicts these processes for a set of related organizations. The figure shows the connections between senior management and one suborganization. Connections to other organizations have been omitted for clarity. Each operation itself consists of multiple suboperations performing tasks directed toward the objectives of the organization as a whole.

A discussion of the meaning of the various boxes, control and instrumentation lines, and other devices shown in the figure is given in the following. VSM is made up of six processes and connected communication channels (see Figure 3.3). Beer referred to these processes by System or S designations:

System 1: Operation or implementation units. This is where the organization produces what the customers/users want. It is the production area where goods, cars, electricity, etc., are produced. Note that S1 management is the local management or supervisors in direct control of the work.

System 2: The coordination activities are here, coordinating activities between production and control/management.

System 3: This is the management and control used to inform production, System 1, what is required of them and to monitor activities.

System 3*: This is an auditing function, which cross-checks that systems S1 and S2 are working effectively.

System 4: This operation looks at the external environment to examine the acceptability of the products and see how the market might change, producing an intelligent prediction of market/environmental changes.

System 5: This is the management policy group balancing the current and future direction of the organization.

The whole concept of VSM is to replicate the ability of organization to respond to changes in a flexible manner. Issues can occur in the environment to problems with production, etc. Rigidly designed management structures do not respond quickly to changes. The other concept pointed out by Beer is to devolve considerable responsibility to the lower level, that is, S1. The closer one is to the action, potentially the faster the response or recovery. This is conditional on having transferred enough authority to the lower levels for them to feel confident to take the necessary actions. Clearly there may be situations that fall outside of S1 fields of competence, in which case the top management, S3, needs to be involved. If S1 operations can respond, then this is good. However, if changes have to be made to the processes themselves or the rules governing the operations, then the management needs to be involved. This may call for budget changes or basic company policy; then these decisions may take a lot of time. Actions involving the government take a long time to occur; one example of this is the Japanese government's response to the tsunami-induced nuclear plant accidents at Fukushima, in March 2011.

## 3.8   VSM APPLICATION TO AIRCRAFT TRAFFIC CONTROL STUDY

This study was the result of PhD research by Al-Ghamdi (2010). Dr. Al-Ghamdi was a student of Dr. Stupples. It has been selected as an example of Beer's VSM as applied to a nonnuclear organization. Some people may consider that VSM is only useful for high-risk organizations (HRO), whereas it can be applied to any organization. The study by Dr. Al-Ghamdi was undertaken to examine the workings of the Saudi ATC to examine its reliability of operation and make recommendations as how it might be improved. This was a mixed problem dealing with both organizational and human reliability issues. For the organizational aspect the study used the VSM approach and for the human reliability issues used the Connectionism Assessment of Human Reliability (CAHR) method of Straeter (2000). The objective of examining this study in detail is to examine how VSM is applied in real studies. This study has a deal of similarity to the study of safety in the case of NPP and other operations, since ATC operations do present issues associated with the reliability of humans and how they can affect safety of operations. Failure of controllers to perform correctly can lead to accidents affecting the lives of passengers, airline staff, and persons on the ground.

A particular advantage of his study was his examination of other organizational methods and also different human reliability assessment (HRA) methods before selecting VSM (Beer, 1979) and CAHR (Straeter, 2000). One organizational approach reviewed was systems theoretic accident model and processes (STAMP) (Leveson, 2004), and Al-Ghamdi decided to use the Beer method in conjunction with CAHR. It is interesting that both Beer and Leveson based their work on cybernetics and referred to Ashby (1956). Leveson's work was also investigated by Spurgin (2013b). In this case, Leveson's STAMP method and its extension system theoretic systems analysis (STPA) in her book (2011a) were examined. Professor Leveson's background is in controls and aeronautics, very similar to the authors. She also considers the cybernetics plays a part in characterizing the whole organization and has considered that computers will play an increasing role in the control of organizations. Our emphasis is a combination of human and computer involvement, with decisions made by humans.

It appears that STAMP integrated cybernetics with human reliability and social influences, but the combination of VSM and CAHR better covered these aspects for Al-Ghamdi's study of Saudi ATC. A review of some of the HRA approaches that were examined is listed later. The use of the Viplan (Espejo, 1989, 1993) approach in the study was a good formalized way of dealing with the requirement to analyze the Saudi air-space incidents. This is another part of the approach that fits well together with VSM and CAHR.

As stated previously, the Viplan approach was used in the study to help organize the analysis of the Saudi air space ATC. It is not proposed to go into details of the approach, but just the steps in the process:

- Establish the identity of the organization (Saudi ATC).
- Model the structure of the organization activities.
- Break apart the structure complexities: establish various structural levels.
- Model the discretionary controls at different levels within the organization.
- Within the organizational structure: study and diagnose the design of the regulatory (control) mechanisms (adaption and cohesion factors).

Quite a number of approaches to solve issues do need a systems approach to break down the structure into separate steps before integrating them to help solve the problem. For example, in the case of studying HRA for inclusion into a Probabilistic Risk Assessment (PRA), there is a method called the Systematic Human Action Reliability Procedure (SHARP) (Hannaman and Spurgin, 1984) that fulfilled that need and was a multistep process. In the case of selecting VSM and CAHR, Al-Ghamdi had the support of Viplan and the CAHR systems analysis structures in helping him in performing the ATC study.

## 3.9  ATC IN SAUDI AIR SPACE

In order to understand the issues associated with this application, one needs to understand how ATC works. Planes are guided from take-off, passing through air space, and landing at their destination. Flights are directed along given pathways (flight paths in space) and hold to different heights, depending on which direction they are traveling. The whole purpose of this approach is to enable freedom of travel of individual planes coupled with a degree of control to enhance safety. So if planes are traveling in the same general air space, safety is ensured by separating the planes by either height or distance. This way, planes should not run into each other, if the pilots follow the rules. This is a virtual equivalent of a highway system. Clearly, planes with different capabilities need to be factored into the mix. Local slow-moving planes should not operate at the same height as fast-moving, intercontinental travel planes, for example, the Concorde operates at heights of 55,000–60,000 ft, well above local commuter planes.

Ground control facilities follow planes traveling in space, approaching to land and landing, and taxiing on the ground. In space, planes are located by radar, and contact with one ground station passes to another as the planes fly overhead. Contact is maintained between various ground stations via changes in frequency of communications. Thus, pilots change the frequency of communication channels as they fly from one ground station to another. Controllers also hand over control in this manner.

As planes approach their proposed landing strip, the control passes to an Area Control Center (ACC), then to the Approach Control Center (APP) and lastly to the Tower (TWR). Figure 3.4 shows all of the air control aspects from far space to ground movements. The process passes the plane from one controller to another controller, so there is contact at all times between the pilot and a controller. It is, of course, not continuous, but only as required. Thus planes are tracked all the way from the distant space to local air space to touchdown and when maneuvering on the ground. ACC covers the contact with pilots from 16,000 ft and above, APP covers contact from 16,000 to 4,000 ft and TWR from 4,000 ft to the ground, including ground movements.

The top layer is the navigation services, which follow planes arriving from foreign air spaces to leaving for foreign air spaces. They also monitor planes within the Saudi air space. This type of process goes on all over the world. Planes are monitored by radar and communicated from time to time as needed by ground sites, as mentioned earlier. Pilots have to register their flight paths ahead of time, and if they want to make changes because of weather, then these have to be revealed and agreed to by the controllers. This is to ensure that planes do not fly into others' flight paths!

The breakdown into various management and control areas is useful, since VSM structure matches the various groupings, and the kinds of actions covered by these various routes affect the reliability of the controllers interacting with the pilots. The author of the study related each of the operations to factors that influence the HRA. So, for example, he related the skills required of the controllers and the tools that they use, such as procedures, displays, etc. He also considered the consequences of errors and whether they are recoverable or not, and what safety indicators are there, and the number of accidents due to this cause, etc.

## 3.10   ANALYSIS OF THE ATM OPERATION

The VSM process diagnoses the organization on various levels and examines the relationships between the management and operational aspects. The power of the Viplan (Espejo, 1989, 1993) was that it helped Al-Ghamdi's understanding of how

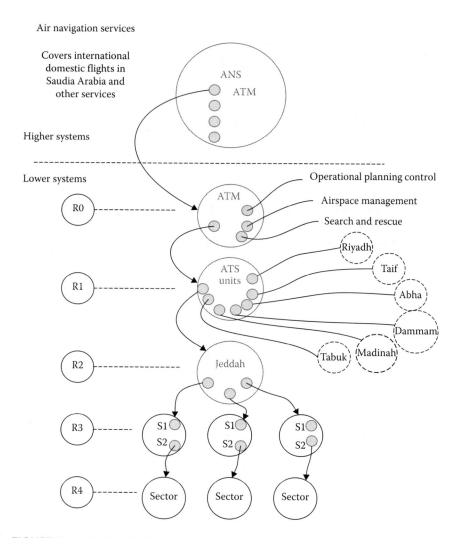

**FIGURE 3.4** Relationship between services areas. (After Al-Ghamdi, S.H., PhD dissertation, City University, London, 2010.)

the air traffic operations, at the various levels, fitted into the VSM formulation. This in turn helped define the interactions between pilots and controllers and helped one understand the time pressures that the controllers worked under. These factors affect the human error probabilities (HEPs) and the significance of the performance-shaping factors (PSFs).

Al-Ghamdi examined the workings of the Saudi air traffic management (ATM), air traffic sector (ATS), Jeddah unit and ACC, APP, and TWR levels based upon the Viplan approach. Figure 3.4 shows the network of parts that makes up the Saudi Arabia ATC system. As can be seen, the network includes other sectors, airports other than Jeddah, but the lower-level elements are associated with Jeddah, and of course, the arrangements for the other airports are very similar.

The VSM figure relating to ATM and Saudi airspace organizations depicted in Figure 3.4 is shown in Figure 3.5. That indicates the relationships between the head of the Jeddah operations, the planning head, and the head responsible for day-to-day operations at Jeddah. The functions are broader than just the commercial ATC functions. Of prime interest in the study were the commercial ATC and the roles of the controllers and pilots as far as reliability of ATC operations. One can see by examining Figure 3.5 that there are managers associated with ACC, APP, and TWR functions. For each level shown in Figure 3.4, a VSM can be constructed which reflects the characteristics of that level. Each of the VSMs is similar in construction, but there are some features that are different; for example, the VSM representing ATS units has seven S1 elements corresponding to each of the individual airports—Jeddah, Riyadh, etc. Figure 3.6 shows a series of icons representing a VSM array corresponding to the elements shown in Figure 3.4. This figure shows a representation of icons, which are unfortunately the same icon, not the actual icons, but the array is correct.

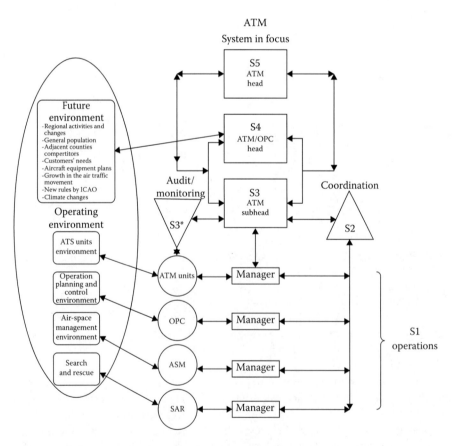

**FIGURE 3.5**  Air traffic management. (After Al-Ghamdi, S.H., PhD dissertation, City University, London, 2010.)

## 3.11   HUMAN RELIABILITY ASSESSMENT

The safety of Saudi air space depends not only on the organization of the air-space control, but also on the reliability of the pilots and controllers to understand and coordinate their activities. The VSM development undertaken by Al-Ghamdi covers the organizational aspects and also covers the different duties carried out by controllers, supervisors, and managers at the different levels within the ATM organization.

A key part of the complete safety and control study of the operation is the assessment of the human reliability probabilities of the staff, as they function in performing their control functions. To that end, Al-Ghamdi evaluated a number of HRA methods and techniques. He selected the CAHR approach by Oliver Straeter (Straeter, 2000).

The CAHR method was developed on the basis of NPP data studies for German nuclear power plants, but has also been applied to ATC studies for Eurocontrol (European air control system). There are several points of interest with respect to CAHR in dealing with PSFs. Straeter considers contribution weighting for each PSF, and a separate weighting of tasks depending on their perceived difficulty. The concept of PSFs started with the father of HRA (Swain and Guttman, 1983) as a method for taking a basic HEP and modifying the basic value to account for the differences between the interaction associated with the basic HEP and the current HEP being evaluated. A large number of HRA studies have used this approach. Straeter also used a modifier based on Rasch (1980) to account for the difficulty of a task being performed by a controller. Easier tasks have higher success values than more cognitively challenging tasks. He also makes use of modified influences of some PSFs in given scenarios. A large number of HRA approaches have used PSF corrections for human reliability numbers, but Straeter seems to be the only one using the Rasch modifier.

Straeter has used a programmed way of dealing with the analysis of events to not only collect errors seen in the event but also to help to identify what PSFs should be used for the given error. The key elements of Straeter's tool are as follows:

- Framework for a structured data collection
- Method for qualitative analysis
- Method for quantitative analysis

It has been stated that CAHR, as used, is a virtual advisor for performing HRA studies related to ATC controllers (Trucco, Leva, and Straeter, 2006). Al-Ghamdi states that some 42 Saudi air-space events are used in the study to evaluate error types and causes of human failures. The analysis of the events shows that there are 309 errors altogether with 262 due to controllers and 47 due to pilots.

## 3.12   LINKING VSM AND CAHR

VSM indicates the arrangements between the management, controllers, and linkages to the planes and pilots as they move through the air space from takeoff to landing. When events occur, they are analyzed, and the method used here is CAHR.

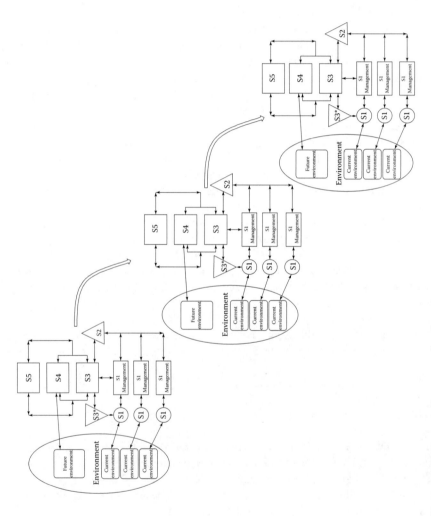

**FIGURE 3.6**  VSM diagrams relating to various operational stages.

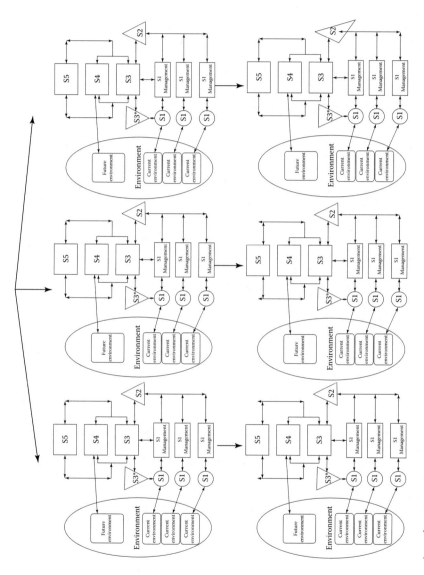

**FIGURE 3.6 (*Continued*)** VSM diagrams relating to various operational stages.

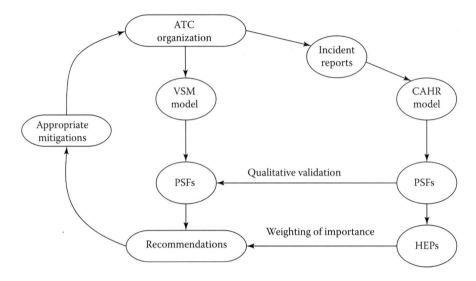

**FIGURE 3.7**    A framework for integration of CAHR and VSM.

The objective of the analysis process is to understand the errors made by the parties and how learning can improve the error rate and reduce the number of incidents. Figure 3.7 depicts the linkage between the VSM models of the ASM process and the CAHR method for evaluation of incident reports, analysis, and recommendations. This is an iterative process with improvements being made to the man–machine interface, training, procedures, and, of course, introducing more controllers to help distribute the controllers' workload.

## 3.13   COMMENTS

The ATC study of Saudi air space is a good illustration of the application of VSM and also shows the advantage of supporting VSM applications with a systematic process like Viplan for directing persons applying the VSM method. The HRA approach taken by the investigator is a good choice for this application in that the method is supported by the help given in the analysis of the accidents, as well as the qualitative method. Often, HRA methods focus mainly on the quantification approach. One can raise some questions about some of the details of the analysis of the HEPs and their application, but this is not the objective of reviewing this study, which was the totality of the application.

The study itself made some recommendations related to the need to increase the number of qualified air traffic controllers based upon the study of the reliability of the current group of controllers. The PSFs covered by the study were ones that were used as part of Straeter's work as part of ATC in Europe (Straeter, 2000). The PSFs covered man–machine and training aspects, which would have been very similar to world standards, so one is left to only improve workload of controllers by requiring more support for controllers by increasing numbers of controllers and by addition of support staff. This is confirmed by Al-Ghamdi's findings.

## 3.14  SUMMARY

The Saudi air-space problem undertaken by Dr. Al-Ghamdi was the examination of the reliability of controller operations in flight control over the Saudi Arabian air space and to make recommendations on how to improve reliability of complete operations. That objective has been met. The study was helped by using VSM, which is a systems approach based upon cybernetics. The concepts of cybernetics have been introduced via its relationship to controls and control theory. Here, the field of control has been illustrated by the introduction of a simple one-loop controller and how it works. In later chapters, we will relate the application of VSM to the field of NPP safety and will build upon the knowledge developed and analyzed here.

The objective of this chapter was to introduce VSM and illustrate its application to a real problem. Two things were done here: (1) introduce the VSM and its relationship to ideas about control systems, and (2) use a real application of VSM coupled with another concept, HRA, but not related to nuclear power. There was a need to show safety could be related to fields other than nuclear. Throughout the book the intent is to illustrate that safety is not purely related to nuclear activities and that management activities can lead to accidents or to the failure to respond correctly. This is just as easily done in the nonnuclear world as it is in the nuclear arena. In fact, some accidents, outside of nuclear, have greater consequences than those that have occurred in nuclear plants. One seems to remember, very clearly, nuclear accidents like Three Mile Island Unit #2, Chernobyl, and Fukushima, but what about nonnuclear accidents like Tenerife (a multiplane crash in which 583 persons were killed and 61 injured) or the Bhopal fertilizer plant disaster in which from 2,000 to 200,000 persons were killed or injured? In fact, the Fukushima-related tsunami killed some 20,000 persons and demolished 140,000 buildings, compared with 2 persons dying at the station. Of course, this is not the whole story: the NPPs were damaged and the reactor cores present long-term cleanup problems. Cleanup problems hold for TMI #2, Chernobyl, and also for Bhopal (chemical: methyl isocyanate [MIC]).

Accidents always have undesirable results: public and staff get killed, plants are destroyed, surroundings are affected, buildings destroyed, and companies go bankrupt. In Dr. Al-Ghamdi's study, the aim was to see how safety and efficiency could be enhanced. The enhancement could be accomplished by more manpower or perhaps by better training.

Our accident studies indicate that much can be achieved by better training of top management personnel. Dr. Al-Ghamdi indicates that other improvements may also be needed. His recommendations have to be acted on by senior managers in his organization to be effective. This makes the basic point that much responsibility for improving safety and the economy lies at the feet of top managers.

# 4 Ashby's Law of Requisite Variety and Its Application

## 4.1 INTRODUCTION

The book is concerned with organizations in the high-risk domain, especially during moments of stress that can occur during accidents. One can separate an organization or business into an operational or production system, a set of control systems, and a set of maintenance operations that keep the production system functioning. See the discussion in Chapter 3 related to Beer's cybernetic model of an organization, which is used here as a dynamic model of an organization.

An operational or production system is one that produces products or supports other organizations/societies. However, the activities of these organizations have to be controlled to ensure that they are operating correctly; that is, they must be stable and produce the required products or services. *Stable* means that the system operates under control and responds to actions demanded of it, with no wild or large deviations.

Ashby's approach deals with relationship of controllers or control systems, including manual actions, to the process being controlled. His work gives us insights into the reasons for the poor behavior of complex systems, resulting from the mismatch of controllers to the systems being controlled.

The basis of his work relates the variety of a controller to the variety of the system being controlled and the consequential stability of the combination. The controller must have a sufficient number of states relative to the system being controlled. These are the number of requisite states to ensure stability. Ashby says that this process is covered by his "law of requisite variety"; that is, the variety of the controller matches the variety of the process.

The selection of the controller or control mechanism depends on the view of an observer of the system in selecting the controller. If his or her view (understanding) of the system is correct, then the selected controller will operate as desired and the system will be stable. However, if his or her view is incorrect, then the system is likely not to be stable.

## 4.2 GENERAL APPROACH TO CONTROL OF SYSTEMS

An understanding of Ashby's law is central to an understanding of how control systems can be designed to work effectively. However, Ashby's book (Ashby, 1956) on cybernetics gives a limited view of the world of both systems and controls.

His book reflects the world of the late 1950s, with the emphasis on relatively uncomplicated mechanical/electric controllers. The systems were complex, but the tools available at that time to analyze systems were limited.

Prior to the advent of computers, there were slide rules, pencil and paper, and some electrical calculators, that is, glorified adding machines; thus there were limits, given current methods to solve these problems. This did not mean that society was intellectually limited. This is shown by Ashby and many others, such as Von Neumann, Shannon, and Lyapunov. This is a world before widespread use of computers and the resulting construction of optimized plants with the need for tight control of variables leading to interactive controller designs. These later computer developments made the need for Ashby's law more necessary. The engineering world situation from circa 1950 and the changes in analytic methods, which occurred in some industries in the following years, indicated the impact of computer technology on the thinking of how to analyze complex systems.

The full applicability of Ashby's law is seen as one moves away from the simple world of the one-loop control system (see Figure 4.1) into the more complex world of multi-loop control systems controlling highly complex industrial systems (see Figure 4.2). The one-loop control system (Figure 4.1) shows a basic control system and uses a single sensor to represent the state of the process. The controller sends a signal to adjust an actuator to compensate for the effect of a disturbance or when an operator changes the set point of the measured variable. The controller settings can be changed in response to a change in the state of the process, either manually or automatically. This kind of controller design can be seen in the case of a power plant, where the process dynamics are a function of the power level of the plant. Similar compensations can be made in the design of aircraft controls to compensate for changes in altitude.

Figure 4.2 shows schematically a nuclear power plant (NPP) and the organization associated with it. This is in contrast to the single-loop controller and system depicted in Figure 4.1. The control systems depicted in the figure are the NPP

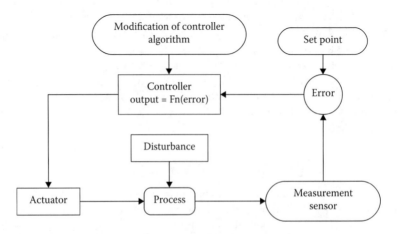

**FIGURE 4.1**   Simple one loop controller and process.

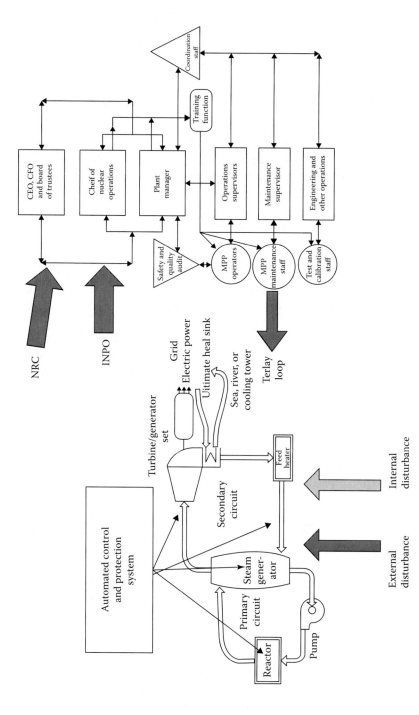

**FIGURE 4.2** Complex organizational control and process.

automatic controls and the reactor protection systems. This latter system acts automatically to safeguard the reactor in the event of a disturbance, which can damage the reactor and plant.

Depicted in the figure is a model of an NPP organization based upon Beer's cybernetic model of an organization (see Chapter 3). Control of an NPP is divided between the automatic control system and a manual control system directed by management and operated by control room staff. Management control over the plant is effected via the control room operators, based on rules, and reinforced by training and education.

Disturbances, which can affect the plant, can be divided into two groups, as depicted on Figure 4.2. These are external disturbances, such as earthquakes, and internal disturbances, such as steam line pipe ruptures. Disturbances can affect the plant and effect changes in the dynamics of the plant, often in a random and unpredictable manner. It means that accidents can change in the requisite variety relative to the normal operating state. Actions taken by the organization or controllers have to account for this variety of changes to be effective in terminating or mitigating the effects of the accident progression. In Chapter 7, a number of accidents and the role of the organization in trying to combat these accidents are considered. Sometimes the accidents are simple events, but other times the accident initiators are not simple, for example, in the case of the tsunami/earthquake at Fukushima (2011).

Figure 4.2 also shows other organizations that can affect the behavior of the organization, namely the regulator, the Nuclear Regulatory Commission (NRC) and the US industry–funded support/review group, the Institute for Nuclear Power Operations (INPO). These organizations are helpful in identifying possible accident causes or improving the operation of NPPs, thus helping to reduce the probability of an accident and its consequence. In other countries, the local regulator and World Association for Nuclear Operators (WANO) are the equivalents to the NRC and INPO.

This book deals with a class of controlled systems, such as NPPs; however, Ashby's law is not limited to just this class of organizations. In fact, Natalia Danilova has applied considerations of Ashby's law to a study based on the web as a data source to research *Search Theories* and then analyze the evidence supplied to determine what information is required to support decisions based upon the investigations (Danilova, 2014). Here the data included in the web has to be defined as relevant and criteria have to be selected to carry out the search, to enable the needed information to be presented to the decision-maker. The process is not as easy as it sounds. There are considerations of uncertainty in the data, and sometimes the decision-maker is not clear on what information is needed to make a decision. Danilova investigated the issues associated with this and found methods to perform the collection and analysis processes to establish quality information.

In the work undertaken here, one is looking for much the same thing: How can one view complex systems to garner the essential information about the system and make the correct decision? One is trying to understand the role of the decision-maker and the relationship of his or her view of the system, as formed by experience, training, and outside sources of advice. This is not just about considering the available data to represent the state of the plant for use by the decision-maker, although that is important. Ashby enables one to perceive that to control the system, one needs to understand the "requisite variety" associated with the system. Understanding a system is not sufficient for a manager, who may be heavily influenced by his or her background. He or she needs to be aware that, during an accident, the "requisite variety" may change, due to damage-induced reconfiguration of the plant and its systems.

When one considers the world of controlled systems, one has to think of all the components of the plant: the plant itself, the controllers, and their interactions. This is particularly so when we consider control systems, made up of combinations of machines and men. In the case of automatic controls, the designer selects a controller design based on his or her concepts and analyses. Before the controller design is fixed, the combination of system (plant) and controllers is tested. However, in the case of persons integrated into the controlling mechanisms, their decisions are affected by considerations of cost, risk, and other judgmental effects, such as the level of experience and knowledge of the behavior of the system (plant). In some cases the real decision-maker maybe a local manager of the operation; so his or her perspectives may condition the actions taken.

Thus, instead of a simple process controller and a system responding to demands, we have a much more complex process, where the responses are conditioned by the characteristics of the human-controller, the person's analysis of the system, and its variability under different operating conditions, including accidents.

We are faced with particular aspects that need to be considered, namely the characteristics of the decision-maker and that of the system and the relationship between them. The characteristics of the decision-maker can shape his or her view of the system and therefore what actions are taken in response to changes/disturbances, which affect the system. This maybe a reaction to the disturbance or predetermined, based on a number reasons, such as the perception of the risk of a future accident. There are examples of cases, some of which are discussed later, where the decision-maker takes or does not take an action based on the probability of a future event. Examples of this are the different actions taken by management before the great tsunami that hit both Fukushima and Onagawa NPPs in Japan.

One has to pay attention not only to the characteristics of decision-makers, but also how one analyzes very large systems. If it is difficult to determine Ashby's law with small systems, it is even more complex in dealing with very large systems. There are often different analytical approaches to deal with large systems; one such way is by reducing the order of the system or by attenuating the system and then later restoring the system to represent reality. The question is how to do this in a way so that one does not lose the essential characteristics of the system.

In dealing with power plant modeling as an example of a large system, one approach was to reduce the contribution of high-frequency contributors. This was done by reducing the linear matrix model of the system by eliminating the higher Eigen values and corresponding vectors. The effect of this was to give the appearance that the response of the plant was accurate, but, unfortunately, removing these contributions could lead to an incorrect estimation of the stability of the system (controller gain increases could not lead instability, which is normally what one expects). The predicted response of the truncated system was always stable, even for high controller gain settings!

The aim of reducing the size of the model of the system was to improve one's capabilities to handle large systems, but also to maintain the correct estimation of system response as required by Ashby's law. This is not an easy task. Clearly, in the previous case, the high-frequency components are required to ensure that controller response is accurately represented. Predictions of the plant's real response are captured and when these effects are needed, one has to include these effects in the mathematical model of the system. For this type of problem, analog or hybrid computers are most appropriate and were used to study this effect. However the arrival of digital computer has led to these machines being phased out. Digital computers, while providing relatively inexpensive solutions, do have difficulties in solving some large-scale plant problems, because of the range of Eigen values associated with different components and their finite element representation (Eyeions, Seyfferth, and Spurgin, 1961)!

## 4.3   IMPACT OF ASHBY'S LAW

The importance of the Ashby's law is related to determining the effectiveness of controls on systems. As such, one needs to understand the dynamics of systems. Failure to understand the system dynamics eventually leads to some kind of failure, since the decisions made by the control mechanism do match the needs of the system; that is, Ashby's law is broken. Later, examples of failures to match the requirements of the law are given.

As mentioned earlier, the view of the designer of the control system with respect to the system to be controlled can determine the efficacy of the process. The decision-maker or designer's understanding of the system establishes the variety of the system and hence the control system efficacy! Of course, in the case of a simple engineering system, one could construct a mathematical model, built on first principles, to determine the variety of the system. However, for complex systems, this option maybe difficult to achieve.

Beer (1979) pointed out that variety of the system could be "destroyed" by managers, making it difficult to have effective control of the process. This destruction may be achieved in a number of ways by a lack of knowledge in the behavior of the system or the philosophy in how the system is expected to respond, which may be wrong! One can see this many times in the political and business world, where a political leader/manager desires a given solution, but this can often not be realized, given the actual dynamics of the system.

## 4.4   EXAMPLES: APPLICATION OF ASHBY'S LAW

A variety of examples have been selected to cover a number of circumstances from different fields in order to show the power of Ashby's law. The examples are from the field of experimental physics, the design of plant equipment, and the cause of a nuclear accident.

### 4.4.1 FERMI'S CHICAGO PILE (CP1) NUCLEAR EXPERIMENT AND XENON

This example shows how an incomplete accounting of the dynamics of nuclear fission processes could lead to a potential failure of an experiment to establish the stable control of the nuclear fission process. Figure 4.3 shows diagrammatically the effect of a neutron hitting a $U_{235}$ atom. The result is that a $U_{235}$ atom splits into two or more neutrons plus so-called fission products. It was this neutron formation that could lead to the cascade process, that is, the ability to grow the number of neutrons under the appropriate conditions, such as in the CP1.

It was Fermi's genius (see Enrico Fermi, atomicarchive.com) that enabled him to come to a conclusion that one of the fission products, namely xenon, had a high capture cross section, which led to the shutdown of the "pile"; that is, the neutron density decreases.

A word about the nuclear experiment, which was carried out at the University of Chicago: The experimenters understood that it was possible theoretically to build a sustained nuclear reaction leading to growth of the number of neutrons. The problem was how to achieve this in practice. They decided to build a reactor out of a pile of graphite blocks and uranium bricks. To start the neutrons flowing, they used a source of neutrons and then observed if the rate of neutrons increased or decreased.

They then progressively added graphite blocks and uranium bricks to build up the size of the pile (reactor), and there was a control rod (neutron absorber) that could be inserted into the pile to ensure that the assembly would not reach criticality unexpectedly (positive increase in number of neutrons).

It is not necessary to cover the whole field of nuclear physics to understand the growth and decay of neutrons within the pile. However, one needs to understand how the original pile experimenters thought of it. They focused on the following equation:

$dn / dt$ = production of neutrons – loss of neutrons through leakage

– loss of neutrons by absorption                                         Equation (4.1)

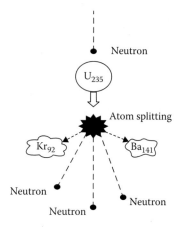

**FIGURE 4.3**   Diagram depicting the splitting of the atom.

Here the absorption of neutrons was considered simply as added "poisons" (absorbers), such as the control material (cadmium) and other impurities in the core fuel or graphite, that is, the effect of poisons on the characteristics of the pile that influence criticality. The core is said to be critical when there is a balance in the production and loss of neutrons. Of course, the balance may be achieved at any net neutron density (or power level). Up to a certain pile size, the criticality of the core is negative; that is, a perturbation in the neutron level (due to the neutron source) would decay.

The rate of decay would decrease as the size and effectiveness of the mixed uranium/graphite pile increased. This process continued until the rate of neutron density turned positive; at this point the control was inserted to add poison and establish a stable neutron density. Thus the understanding of the experimenters matched the behavior of the reactor system.

However, once the reactor reached such a stable condition, the neutron density started to slowly decrease at first, and the control rods were withdrawn to bring the reactor to a stable condition. Eventually the control rod was completely withdrawn and the neutron density continued to drop. Thus, the actual behavior did not match the experimenter's model of the reactor process.

The previous description illustrates the contention about the relationship of the observer's view of a process and the actual variety of the process. Unless the view of the observer and the process match as far as the variety is concerned, the control of the process can be affected.

In this case, Fermi developed a view of the complete neutron process, including the accounting of the influence of the decay products resulting from the fission process. In the fission of the uranium, not only were a number of neutrons ($\sim >2$) produced, but also other products, including iodine$_{135}$, which decays into $X_{135}$, and this was the element that Fermi decided had a very large cross section (high probability of capturing neutrons).

Fermi realized the impact of xenon on the behavior of neutron density and the need to compensate for the effect of xenon, an absorber of neutrons. An early model of the fission process was modified, and thus he became aware of the real variety of the process and therefore was able to more effectively control the process and ensure stability of the neutron density by increasing the size of the pile and controlling the impact of xenon by rod movements—that is, balance the growth of one poison by the removal of another.

Sometimes, in practice, one is not able to recover from the failure to predict the effect of not correctly identifying the variety of the system that one wishes to control. A step taken by some to compensate for this lack of understanding of performance is to build a simulation of a complex system, in order to try to reach beyond the limits of a designer's view of the states in a process, that is, its variety. However, building a simulation of the process, one needs to include all of its variety, but in a simulation it is easier in that one can capture the variety of smaller units and integrate them to form the complete plant simulation.

### 4.4.2 Effect of Management Decisions on the Fukushima Accident Progression

The Fukushima tsunami-induced accident, which occurred in Japan in March 2011, was the worst accident to affect a set of NPPs in recent years. The organization in charge of the plants was Tokyo Electric Power Company (TEPCO). The NPPs affected by the accident were of the *Boiling Water Reactor,* a type designed by General Electric (US) and Toshiba (Japanese) under license.

The primary issue was the failure of TEPCO management to deal with the possibility of an earthquake-induced tsunami, which could exceed the height of protective seawalls and could lead to the flooding of the reactor area, preventing the emergency power (diesels) or the switchgear from working and stopping the instruments operating key electrical protective and control circuits.

In the other failure, which is dependent failure, they failed to plan for what actions that the station staff should take and what equipment might be needed if the staff had to respond to a large tsunami. The details of the accident are covered in Chapter 7 and are not repeated here, beyond the fact that the TEPCO management was unprepared for the accident.

The point to be made here with respect to the application of Ashby's law is the fact that the variety of a system should consider the possibility of an external influence, which could modify or change the variety of the system. The view of the system, including the effect of outside influences, has to be considered by the observer/decision-maker before the correct means of control can effected. In the case of TEPCO, the management appears to have decided that a tsunami of the magnitude that occurred was not likely to occur, and they chose to mentally block consideration of the possibility of a 45 ft tsunami occurring. Thus the ability to consider the possible consequences was ignored.

This brings us back to the statement of Beer (1975) on the fact that management often destroys the variety of the system to fit its needs at a given time. It is believed that TEPCO did not wish to acknowledge the possibility of a large tsunami, since they would have to shut down the reactors and suffer a loss of revenue until the bigger seawall was constructed, plus the fact that it would cost a large sum of money to build. So their concern over costs overrode advice given on the possibility of a large tsunami and the big loss for TEPCO that could occur and also that the reactor core damage could impact the safety of the surrounding population.

This is a case where the management did not want to accept the risk associated with a large tsunami and decided that the predicted low probability of the event was a way of avoiding making an uncomfortable decision. A superficial review of the effect of a large tsunami on the plant could lead one to fail to realize the total consequences, since the effect of the tsunami affected the plant in a large number of ways beyond just the loss of power. It effectively destroyed diverse ways of recovering the plant via the actions of station staff. The plant characteristics under severe accidents were not understood by the decision-makers. This underlines the necessity of being aware of Ashby's law, even under post-accident conditions, as well as the interpretation of probabilities.

### 4.4.3 Failure of San Onofre NPP Steam Generators

Another example of the application of Ashby's law is one related to the decisions taken to replace the aging steam generators at an NPP in Southern California called the San Onofre NPP. Here the impact of Ashby's law is related to the impact of management's lack of knowledge about steam generator performance and the impact of steam generator modifications on performance (see notes contained in Joksimovich and Spurgin, 2014).

The lifetime of an NPP's steam generators is limited to about 20 years, which is less than the expected lifetime of the plant, which is 30 to 40 years. Therefore, it is necessary to plan to replace the steam generators with new units. This has been done successfully many times for a variety of nuclear stations throughout the world, and the lifetime of the new units is usually better than that of the original steam generators.

The life of steam generators can be affected by a number of different mechanisms. These mechanisms are corrosion, stress cracking, and vibration-induced tube failures. Over the years these mechanisms have studied by a number of organizations, including the Electric Power Research Institute (EPRI). As a result of these investigations, methods have evolved to mitigate all the effects, including that of fluid-induced vibrations. These fluid flow–induced vibrations can lead to tube wear and subsequent release of radioactivity, which is a safety issue.

The principle owner of the San Onofre NPPs selected the Japanese company Mitsubishi to build the replacement steam generators. However, the steam generators were not replicas of the original designs built by Combustion Engineering, which was the designer of the plant. There were considerable differences in the replacement steam generators' internal structures compared to the original units, and this should have led to some concerns about their performance and life.

The replacement steam generators have lasted less than three years. It is not clear how the decision taken by the parties has led to this state, since the internal discussions and decisions have not been released and are likely never to be! Ultimately, the decision must rest at the feet of the top management of Southern California Edison (SCE). It is they who approved the choice of the manufacturers of the replacement steam generators and the design modifications to the steam generators, along with the selection of their designers.

There are a few observations to be made before forming an opinion on the effectiveness of the management decision-making. It should be clear that management needed a good understanding of the various issues that could strongly affect steam generator life. Failing that capability, the decision-makers should have had competent advisers to help them formulate the specification of the steam generator design. Judged by the result, it appears that neither of these conditions were met.

The only information available to the authors was a paper written jointly by persons from SCE and Mitsubishi. It focused on quality control rather than the very important issue of tube vibration during operation. Equally surprising was the failure of the SCE/Mitsubishi design group to seek advice from EPRI, especially since SCE had access to EPRI's research work.

So how does this relate to Ashby's law? As was mentioned previously about the role of the decision-maker as an observer looking at a system, if the observer's view is incorrect, as far as the variety (states), then his or her ability to be aware of what is required is severely impaired. It is then impossible to design a controller/control scheme to perform as required. In the case of steam generators, the control of vibrations is passive. The design to control motion is by the use of suppression bars to control the movement of the tubes during all phases of operation.

So this example is another view of the application of Ashby's law. The key is the relationship of the observer view of a system and the reality of the system. Beer talked about the destruction of variety of a system by managers. The destruction of variety maybe achieved by actions or failing to act—driven by ignorance, poor advice, and deliberate choice (driven by ideology)!

## 4.5 METHODS TO ENHANCE THE PROBABILITY OF GOOD DECISION-MAKING

From the previous examples, one can appreciate the need on behalf of the decision-maker to enhance his or her understanding of the system about which decisions are being made. It is also clear that despite improvements in the decision-making processes and aids, at times the wrong decision will be made. The objective is to improve the process and reduce the probability of a mistake.

The general approach is to first appreciate that something must change in how decisions are made. The ability of decision-makers to improve their capabilities rests on improved training, education, and the selection of capable support staff. An understanding of Ashby's law and its applicability to evaluate situations involving systems of different kinds should be part of the training process.

## 4.6 CONCLUSIONS

One can see from the previous discussion that an understanding of Ashby's law and its application is extremely important and an essential element in good management decision-making, especially in high-risk situations. Not only does one need to define the limits of understanding of a system, but one also has to be aware of the limits of the observer/decision-maker to correctly perceive the key states/variety of the system and how they interact. It should be pointed out that systems involving humans are much more difficult to deal with than mechanical/electrical systems that can be closely defined and bounded by mathematics. This presents a need for the decision-maker to be advised by competent and experienced individuals, but even with good advice, one can only reduce the probability of an accident.

The capability of management to make good decisions depends on a number of things, some of which are within the compass of the individual manager, that is, training, experience, and knowledge; also the manager has to depend on the advice of others, usually selected by the manager. Outside of that is knowing the probability of outside and inside disturbance events, so the manager has to have a good awareness of risk/benefit stemming from dealing with these events. In the case of large complex systems, a further aspect of uncertainty is raised in that the manager may not be able to gain a clear image of the state of a system (see Danilov, 2014).

A poor assessment of the risk of operation could lead to the demise of an organization. An example of this is the decision made by TEPCO management in the case of the Fukushima tsunami (see Chapter 7), which led to the near demise of TEPCO. Essentially, the manager has to be aware of Ashby's law and its implication in terms of what needs to be understood about a system. Of course, accidents can effectively change the manager's perception of the behavior of the system. To combat this, the manager needs to pay attention ahead of the possibility of an accident and its progression, especially if it could severely impact the operation of the organization and the configuration of the plant. In the case of the Fukushima accident, the impact of secondary earthquakes and damage caused by the tsunamis defeated the efforts of the plant manager and crews in trying to recover from the effects of the initial tsunami (see Chapter 7).

# 5 Probability Risk Assessment

## 5.1 INTRODUCTION TO PRA

It has been suggested that management operations in the field of safety and economics can be improved by the use of various technological methods and procedures; among those tools is the use of probability risk assessments (or safety assessments) (PRAs). This chapter briefly covers what a PRA is, its foundation, and its use by management as part of a support system to assist management. PRA could assist in making judgments about some of the risks that could affect the operation and their estimated probability (see, e.g., Frank, 2008). PRA can be a valuable tool, when used effectively. Of itself, it is not without its limitations, such as: What is the justification for the failure probability estimates of equipment and persons and probabilities of accident initiating events (IEs)?

PRA methods were started by the US Nuclear Regulatory Commission (NRC) as a method for better understanding the risk of nuclear operations and then using it to shape their response to what to focus on and what could be of lesser risk. The methodology grew out of the fault tree methodology used by NASA to examine how faults could lead to rocket/mission failure. The fault tree methodology seemed to be focused on equipment failures. However, human errors related to maintenance operations can be incorporated into the fault tree logic along with equipment reliability issues. This can be done in a PRA by separating the system equipment failure data from the human errors, rather than lumping everything under equipment availability numbers. Which approach is used depends upon the record keeping of the organization on the root causes of equipment failure.

The NRC group under the leadership of Professor Rasmussen (MIT) and S. Levin (USNRC) (WASH 1400, 1975) developed the PRA method and based it upon of the idea of an event tree. This is a method of logically connecting equipment failures with human errors and IEs that lead to the consequence resulting from this train of events.

One can appreciate that an environment event could lead to some form of damage to a plant, but the situation could be worse, if at the same time there was a failure of equipment coupled with a failure of station personnel to take action to stop the progression of the accident. A basic event tree consisting of an IE, a system failure (mechanical/electrical) (ME), and a human error (HE) failure to operate a mitigating system leads to an undesirable consequence (core damage and release of radioactivity). For each element, analysts performing a PRA study would estimate the probability of the elements and combine them to produce an overall probability of such an accident occurring.

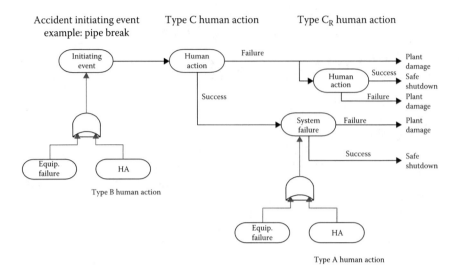

**FIGURE 5.1**  Human events in a PSA logic tree.

An event tree can cover a number of similar branches, involving different events and associated human actions/system failures. Clearly, the tree shown in Figure 5.1 is one that has an IE related to an equipment failure (pipe break) and a human error. An IE may be something like a small-break loss of coolant accident (SBLOCA), a pipe failure due to, say, corrosion, which occurred but should have been detected and prevented by human action (ultrasonic pipe examination)! In the figure, different types of human actions (HAs) are identified. For more information about human reliability assessment, see Spurgin (2009).

It should be pointed out that having a periodic inspection program is the responsiblity of the plant management, to be either technically competent and/or advised by technical advisors to ensure the safe and economic operation of any high-risk organizational (HRO) plant, not just nuclear power plants (NPPs). The Flixborough accident in the United Kingdom was a case involving a chemical plant, where the UK enquiry found that the managers were chemical engineers and lacked the necessary mechanical engineering knowledge related to preventing the accident occurring. It should not be assumed that managers are technically omniscient.

Also, it should be pointed out that particular care needs to be taken during "workaround" operations. An example of problems that can occur is the case of the Flixborough Nypro chemical plant accident in June 1974 (Flixborough, 1974), in which pipe installations failed (different theories whether it was an 8″ or a 20″ pipe failure that was the cause of the explosion; 28 persons killed and 36 injured and much other damage occurred).

A complete PRA study would consist of many different IEs. A recent Electric Power Research Institute (EPRI) report (Sursock and Lewis, 2015) considered the aggregation of risk for *Risk Informed* decision-making. The NRC in its safety assessments use PRA risk assessment as part of its approach to ensuring the safety of NPPs along with other aspects. As far as PRA is concerned (see NRC fact sheet on

Probability Risk Assessment [PRA]), the NRC does not limit its safety assessment to PRA. This methodology is just a part of the approach they call "risk informed safety assessment" of operating plants.

## 5.2   PRA STRUCTURE

In the previous section, a single event is depicted. However, in a full PRA, a number of causal events are considered, so one might consider a number of IEs—strong winds, floods (internal and external), earthquakes with and without tsunamis, loss of off-site power, loss of diesels, etc. All of the these events are coupled with a variety of mechanical components, systems designed to shut down the reactivity of the reactor, such as boron delivery pumps, systems to cool the reactor and remove heat from the core, steam generators, and other components and, of course, the corresponding human errors are considered.

The overall logic structure of a PRA consists of various logical structures combined into event trees as mentioned previously. The model of an equipment system may be combined into a fault tree in which equipment failures and human events (maintenance operations) are represented. The details of the modeling depend on the analyst and the needs of the utility or organization. The degree of complexity of the model depends on the availability of data related to records maintained by the organization. Equipment failure data is much easier to come by; HE data is much more difficult to gather and estimate (see Spurgin, 2009).

It is usual for organizations to examine the safety of the operation under different domains, such as full power, low power, and shutdown conditions. In general, things are likely to happen much quicker at the higher power levels, but different problems may occur under the other conditions, say, fewer persons on watch. Of course, there may be off-normal conditions to examine, such as what occurred at Paks and Barsebeck, as covered in Chapter 7 (Section 7.9).

## 5.3   APPLICATIONS OF PRA

Some of the applications, such as variations in power level, have been mentioned previously. One of the main uses of the PRA is to convince the regulatory authorities that the current state of safety of the plant is acceptable, as far as the public is concerned. Each organization has to ensure that the state of the plant and how it is being operated are in compliance with the PRA and the "Final Safety Assessment Report." The situation with Northeast Utilities (NU), as reported in Chapter 7, indicates sometimes compliance does not occur and can end with the NRC taking drastic actions to change the situation, such as the shutdown of NU's NPPs (see Section 7.9.4).

The value of the PSA is as a tool to help management be aware of how the safety of the plant might be affected by changes in operation, such as taking safety system channels out of operation for maintenance work, before having to shut down the plant. Usually, the ground rules to do this are agreed on with NRC ahead of time. This is part of the so-called technical specification for operating the plant.

Other organizations may not be regulated like the nuclear energy field, but they could use PRAs in a similar manner to estimate either safety or economic effects of modifying the control and protective systems. For some organizations, changes in the consequences could occur if the design is used in a different manner. A case comes to mind with the Piper Alpha oil rig accident, June 7, 1988 (see Spurgin, 2009; Section 8.8 for details). The basic rig was designed for oil operations and its operator safety protection was compromised in making the shift, since the characteristic of going from oil to gas is that gas can lead to explosions, unlike oil, which is more likely to lead to fires. Performing a PRA anew should have revealed the effect of change and the organization should have installed changes in worker protection. Another issue was the failure to protect divers working on the rig when pumps were turned on to respond to the gas explosions—a case of poor safety awareness by the operating organization.

The nuclear industry technology is fairly well understood, but variations can occur and should equally be understood by management as part of their knowledge/training. Sometimes, environment effects are not as well understood and considered and their effects can be underestimated, as they were in the Fukushima accident.

This points to one of the problems associated with the use of PRAs: many people believe that if the probability is very low, the event will not happen in the next few years. As seen in the case of the Fukushima accident, a tsunami of this magnitude had not occurred since year 843, and the estimated probability of similar tsunami event was assumed to be about 1 in 1,000 years. Therefore, one could assume this would not happen soon. Wrong! It occurred, and Tokyo Electric Power Company (TEPCO) was completely unprepared to deal with the size of the tsunami. Their failure covered both the failure to build an appropriate-sized seawall and the inability to respond quickly to the consequences of the flooding on the plant facilities.

PRAs are used to develop the details of operations and are probability related. This gives us the relative information about the variability of risk depending on different events; that is, some are low risk and others are high risk. However, PRAs are not predictions; they estimate the risk of a given situation. It is up to the management to decide if the risk is worth taking. Management has to evaluate the risk–benefit ratio. One of the problems is the estimation of risk as the combination of consequence multiplied by the associated probability. If the probability is very low and the consequence is very high, the risk may seem to be tolerable. Whereas for a company, if an accident does occur, the impact of the accident is such that the cost is so high, that company goes bankrupt and it ceases to function (TEPCO must have been close to this condition, following the Fukushima accident)!

## 5.4  SUMMARY

A PRA is a valuable tool to inform management and the NRC as to the safety state of the plant. Operations carried on the plant can change the safety profile of the plant and change the estimated risk of the operation. Management has to react to the change and take action—by reducing power or other actions. The dynamics of a PRA can change depending on the actions or lack of action by management and staff. Some parts of the PRA have been automated to advise operating staff that a change in the increment risk has occurred due to a safety system failing to operate and that the control-room crew should take action to keep the plant operating within the limits covered in the plant technical specifications.

As mentioned previously, PRA results are not predictions of what might happen, but assessments of the relative risks of events. Any given event within the PRA study could be seen as low risk and happen the next day. However, if the consequence of an accident is low and this is why the assessed risk is low, then the accident may be seen as a nuisance. However, if the risk is low, but the consequence is high, then the matter depends on the manager and how likely he or she thinks that event will occur.

We have comparisons of manager's choices with what happened at the Fukushima NPPs and the Onagawa NPP. The earthquake event was the same for both; the ground accelerations were slightly higher for the Onagawa site. The resulting tsunamis affected both plants, but the Onagawa plant was better prepared, because of the decision of an Onagawa manager to build a high-enough seawall to prevent significant damage to that NPP. The TEPCO managers decided not to build the seawall to a sufficient height and the result was the destruction of NPPs. This accident is discussed in some detail in Chapter 7.

The NASA *Challenger* accident is a similar case in that NASA managers decided to launch, against the advice of an engineer about the effect of cold temperature on the launch vehicle and solid rockets, specifically affecting a set of "O" rings. This accident is also covered in Chapter 7, along with other examples of the effect of managerial decisions on accident progression.

The conservative decision has consequences that later turn out to be the better way to go. What is the loss in launching later, or is the cost involved in building a better seawall so much more than causing the bankruptcy of the company? Were the cost/benefits really looked at carefully?

# 6 Rasmussen's Human Behavior Groups

## 6.1 INTRODUCTION TO SKILL-, RULE-, AND KNOWLEDGE-BASED BEHAVIOR

The role that managers and operators can play in accidents is related to their intrinsic behavior, which was investigated by Jens Rasmussen. He defined the behavior of humans by different classifications called skill-, rule-, and knowledge-based behavior (Rasmussen, 1979). The behaviors reflect how humans respond to a given task based on their prior training. The reliability of their actions depends on the mode of response that they use, so, in general, skill- and rule-based behaviors are likely to have greater reliability than knowledge-based behavior. Also, the speed of response is usually faster for skill- and rule-based modes; more time is taken by persons operating in the knowledge-based mode to understand and react to a situation.

How successful persons dealt with situations depended on their mental preparedness for the situation. Given that persons do respond differently, the question arises of how should management and operators be prepared to respond to both the likelihood of an accident and its progression. The approach developed following the Three Mile Island Unit #2 (TMI #2 ) accident was that control-room operators should be trained to deal with different accidents by being exposed to a series of simulated accidents, with their response based on essentially an approach designed by experts and incorporated in a set of emergency operating procedures (EOPs). In terms of Rasmussen, the operators were trained to respond in a rule-based manner. In addition, their knowledge base was extended by classroom training in nuclear energy, plant dynamics, and nuclear physics. This meant that they could operate in a knowledge-based manner, if the procedures did not seem to work.

Earlier, nuclear operator training was not like this. Operators were given some background training in nuclear and plant characteristics, but the role assigned to them was basically to monitor the actions taken by the automatic control and protection systems and ensure that the instrumentation used by the protection systems was working correctly. In the period prior to the accident, there were very few plant simulators and consequentially few operators were exposed to use of this tool. Also, simulator accident scenarios were limited to just a few accidents based on those used to design the safety/protection systems. In other words, the safe operation of the nuclear power plants (NPPs) depended very much on the capability of the protection system designers rather than the skill of the operators. One of the weaknesses of the old approach was to rely on the capability of the operators to determine the kind of accident from the symptoms displayed by the accident. This was not an easy task since the actual accidents may not follow exactly the symptoms identified by the analysts. The design of the procedures was later changed to respond to the accident symptoms rather than the supposed accident event; that is, procedures changed from event-based to symptom-based procedures.

To understand the differences among the Rasmussen-defined behavior groups is to start with the job of making furniture. Most people think that they can understand how to do this. The woodworker can perform the job quickly and efficiently, but the unskilled person takes a long time to make the article and it may not fit together very well.

So the woodworker is operating in a skill-based mode, whereas the other person is operating in either a rule-based or knowledge-based mode. If the person is in the rule-based mode, he or she can follow plans describing what should be done, but it may be difficult to produce a good finished job. If a person is operating in a knowledge-based mode, he or she relies on his or her knowledge to produce drawings and then make the finished product.

One can see the differences between the three different behaviors:

1. The skill-based person can produce a good job in a timely sense provided that it is something that he or she is practiced in performing.
2. The rule-based person can perform adequately, if one has plans to follow, but to enhance one's performance, one needs practice.
3. The knowledge-based person has to develop plans; one has to then make the product, which may not work well, and one is likely to spend a lot of time modifying and changing the final product.

A skill-based person has to spend a lot of time learning and honing skills. Often, persons become apprentices to learn these skills, under the guidance of a skilled practitioner. The skills need to be practiced often and since one spends such a lot of time focusing on this particular skill, it is likely the person's range of skills is limited. A typical occupation might be a furniture maker. Persons working in an organization like a nuclear plant are not likely to be skill-based, except for a small number of artisans. It is not very cost effective to have many persons like this at a plant. The station needs generalists to perform a number of tasks using procedures: maintenance personnel operating on a range of different equipment in subgroups of electricians, mechanical engineers, etc.

In the control of the plant, a rule-based operation is one in which the operator responds to a given situation by firstly determining the situation from a set of indications (symptoms) displayed by the control-room instruments and then following the rules covered in a procedure to respond to the situation. In the nuclear field, there are groups of procedures—normal procedures, abnormal procedures, and emergency procedures. The procedure selected by the operator depends on the situation. So, for instance, a normal procedure is used to carry out operations like changing the power output of the station. The abnormal procedure is used by the operator to respond to something not working correctly, like sensor failure. The emergency procedure is used when an event occurs that could hazard the plant and possibly affect the public.

Knowledge-based behavior is one when a person is placed in a position to use his or her knowledge to work out how to respond to an unknown situation. Of course, one's knowledge has to encompass the situation, but the situation arises in a way that one is not prepared for. The solution to the problem is not going to be arrived at quickly, since the process of hypothesizing, testing, and confirming takes time. In fact, the accident progression may advance and even change the basis of the examination.

Although the topic is within the skill set of the person, time is not on his side! Hence, using generalized knowledge as the basis for terminating or mitigating an accident is not the best way to go. It was found that the early Russian approach to operator selection was to select highly educated persons to combat accidents; that is, operating in knowledge-based mode. The American way was to rely more on the design of good procedures and use trained personnel with high school education. This approach evolved after the TMI #2 accident. The highly educated personnel are better used in designing the procedures and testing them over a range of simulated accidents to ensure that they work!

## 6.2  APPLICATION OF SKILL, RULE, KNOWLEDGE BEHAVIOR RULES

Why should one be interested in skill-, rule-, and knowledge-based (SRK) behaviors? If one understands the advantages and limitations of human behavior, one can match the requirements to control accidents to the capabilities of the organization. The organizational dynamics are represented by a Beer cybernetic model (see Chapter 4), with policy and guidance being determined by the upper management. The middle management performs two functions—advises upper management and ensures the requirements of the upper management are transmitted to the operators. Operators take the required actions following the dictates of the procedures.

A development that has occurred in the nuclear industry is to appoint someone to better advise the management in nuclear technical matters. This person is appointed as the chief nuclear officer. This person advises the chief executive officer (CEO) in matters of nuclear technology and enhances the capability of the CEO to make good decisions vis-à-vis nuclear issues, such as design of procedures and risks associated with various accidents.

One can appreciate that the top manager operates in a knowledge-based mode, so it takes him or her time to formulate a policy with regard to a set of accidents, especially when he or she has to rely on others. The middle managers also function in a knowledge-based manner, but are likely to have been operators and remember some of the needs and experiences of operators. The middle managers are likely to work with procedure–design experts to develop the procedures where the scope of the activity is defined by the CEO. The procedures are built and tested with aid of the operators. The operators are aware of the physical limitations associated with a new set of procedures. The technical issues with procedures are tested using full scope simulators.

The role of Ashby's law (see Chapter 4) has not been mentioned here before, but one can see that insight is needed into accidents that progress beyond the model basis of the simulator in order to understand how to control the effects of an accident that does not act like the simulation. An example of this is the actions that took place during the Fukushima accident. The approach taken by the supervisor seemed to be guided by a good understanding of the plant but was undermined by subsequent actions effected by subsequent earthquakes and tsunamis, which changed the configuration of the plant and availability of equipment to be used to respond to the accident progression. These configuration changes affected access to control valves and sensor information. The degree of complexity increased beyond the normal procedural assistance and into the world of knowledge-based behavior.

The plant supervisor had to reimagine the plant systems and their controls. That is, he had to see how the requisite variety of the plant had changed, as a result of the accident and its progression. The continuing actions of the earthquake, ground accelerations, and the series of tsunami waves had changed the plant from its initial state. The changes in the requisite variety were occurring through most of the accident period. Hydrogen explosions were occurring, plant configuration was changing, instruments were destroyed or made inaccessible, and actions taken by the staff were negated.

It is difficult to see how the supervisor could really succeed, since he did not have time to reevaluate the whole effect. On top of that, the system was dynamically changing. It is not sure that things he was attempting to do could be done, since things changed as he was trying to do them. The supervisor was faced with a difficult task, and it was made more difficult by having to deal with Tokyo Electric Power Company (TEPCO) executives in Tokyo! Part of the problem was dealing with the consequences of not releasing hydrogen gas to the atmosphere. This could lead to explosions with unknown effects; best to disperse this gas—yet another variable in the set that the supervisor had to deal with!

It was commented that it is better for operators to deal with accidents in the rule-based mode. To accomplish this, one needs to be able to define the set of accidents in sufficient detail to define the set of rules. The approach breaks down if the rules cannot be defined. In the case of the Fukushima accident, a number of prior design decisions were made, which led to increased vulnerability of the plants, variability of the accident sequence, and the availability of tools for the supervisor to use during the accident progression. Clearly, the supervisor tried to mitigate the effects of the tsunami, but ultimately it was beyond his best efforts to do so. It appeared that when he fixed one thing, the accident progressed faster than did his actions.

## 6.3   COMMENTS

This chapter describes Rasmussen's contribution to understanding human behavior in the form of SRK-based models. It describes the utility of the approach in supporting the best way to use humans in responding to accidents. The approach justifies the use of procedures for operators to control the progression of accidents. However, this approach is valid as long as the designers and organizations are capable of characterizing accidents to enable a set rule to be generated. Once one moves away from this state, the crews are forced into a knowledge-based mode, with failure to cope being a strong possibility.

The big advantage of using procedures is that it enhances the success of the operators in terminating a normally expected accident or mitigating its consequences.

# 7 Case Studies of Accidents for Different Industries

## 7.1 SCOPE: ANALYSIS OF ACCIDENTS

The study of accidents can serve a number of useful purposes, leading to enhancing both the safety and economics of industries and organizations. These are (1) the study of specific accidents can help the industry reconsider the safety and cost implications of accidents; (2) the study of accidents by managers can help them understand how accidents can stem from a lack of attention to details and that natural phenomena can override safety guides and protective barriers and lead to an accident that can risk the company; (3) industry regulators can formulate rules and guidelines for the industry to help transfer the lessons learned related to a given accident to the rest of the industry—this is the standard approach for any regulator's response to an accident associated with the industry, and. (4) another benefit that can result from the analysis of any industrial accident is a deeper understanding of the causes that might prevent a similar accident in one's organization. One issue related to accidents is that organizations tend to look at the more superficial causes of accidents rather than looking for more fundamental causes and then drawing the appropriate lessons for their organization.

Lessons learned in one industry can be applicable to another. For example, management techniques can be a common cause that affects the potential for similar accidents to occur. For example, see the following discussion related to Northeast Utilities (Section 7.9.4). Sometimes decisions made by management are determined by financial considerations, that is, things like shareholder value (see discussion in Chapter 13). This seemed to be case for Northeast Utilities: safety was not their prime consideration or even the second consideration!

## 7.2 ACCIDENTS: ANALYSIS APPROACH

Often analyses of accidents follow specific pathways starting with a blow-by-blow history of the accident sequence and ending with consequences. Here we will cover the overall characteristics of an accident and then summarize our perception of the main causes of the accident. This way, a number of accidents can be documented without taking a lot of space and reader time. One can focus on consideration of the findings and draw conclusions on what could have been done to avoid the accident. Of course, this is very presumptuous, but like this, people can be stimulated into thinking, not just accepting the standard answers and solutions. There is a tendency to focus on the immediate solution and on the personnel directly involved in the accident, without thinking about that the situation might have larger implications. For example, Japan has been exposed to both earthquakes and tsunamis. These can be connected events, but the national response seems to be different. Response to earthquakes seems to focus on improving an individual building's response capability, whereas there does not appear to be a global response to tsunamis in Japan, which would seem to be needed!

The process is to identify the possible direct and indirect causes of an accident, the sequence of events, and physical and organizational steps taken, and then analyze the accident in terms of direct and indirect responsibilities. That is, was it caused by a natural event, and what role did operators and management take? Also, were there unexpected equipment failures that modified the effects of the accident? The description of the accident may be expanded to include an event sequence diagram (ESD) to show other pathways that could lead to other different consequences, so that the actual accident pathway may not be exclusive or dominant. For example, the tsunami accident at Fukushima was not just the result of the size of tsunami, but the result of faulty thinking at both the Tokyo Electric Power Company (TEPCO) management level and also the Japanese government!

The idea is to supply readers with a sense of discovery behind each of the accidents and not overwhelm them with details. There appears to be a tendency to cover reports with data without pointing out the significance of the pieces of data. Some of the data are germane and others have no significance. These are our views, and others may have different views and concepts, but one can go from here and make progress, which is always our aim.

Many accident reports focus on reporting things from the bottom up, that is, focusing on the operators, since they are persons that take direct actions, rather than the totality of the situation. One should ask how the organization got into that state. One needs to think from both bottom up and top down. This means not only should one look at the actions of operators at the "sharp end" of actions, but also at the "blunt end" related to prior decisions made by the management.

Operators do not make policy decisions, which are up to management. Decisions can lead to accidents! For example, the decision to build or not build a seawall taken by management in the cases of Onagawa and Fukushima nuclear plants: Onagawa was saved and the Fukushima plants were destroyed. Management decisions can lead to either a good or bad result!

The approach taken here is to break down accidents from initial decision states to accident occurrence and to add reactions to comments on post-accident analyses by other organizations. For example, the Three Mile Island (TMI) Unit # 2 pressurized water reactor (PWR) accident (March 1979) evoked a number of different analyses and even caused the then president of the United States (President Carter) to call for a deep analysis of the accident, because of its importance to the nuclear energy industry and the safety of this form of power generation, as far as the public was concerned.

It should be mentioned that there is a universal love–hate relationship over the world for this type of energy generation, primarily due to the use of atomic bombs in Japan during World War II. A similar response of late has developed to the other main energy producer, based on burning fossil fuels (coal and gas). It should be pointed out that modern society is very dependent on cheap power and the competitive position of one country versus another one rests on the price of electricity. The purpose of this book is not to defend various energy sources, but to look at how they are operated and the consequences of not doing so correctly and carefully! The country should not decide on the basis of some poor actors to change from using efficient and cost-effective power sources.

## 7.3  LIST OF ACCIDENTS

Accidents have occurred in different industries and a selection of some of these accidents illustrates several points that are in common (see Table 7.1). It is usual to focus on major accidents, but it is also useful to consider the characteristics of some of the smaller accidents or even near accidents that have occurred. So the divisions considered here are a set of the major accidents/incidents and a group of ancillary safety-related incidents. The major group of accidents has resulted in high consequences in terms of number of deaths and/or property damage. The ancillary group relates to things that have happened, such as radiation exposure of a few persons due to an incorrectly carried out operation, the impact due to an equipment-design failure and its effect, or the effect of management change on station operations.

Here the objective is to illustrate the role of management in dealing with the accidents/incidents. Even the ones not selected have interesting features and they could have easily been selected to analyze their organizational behavior. Most accidents seem to reveal similar management shortcomings.

The impact of accidents on society seems to depend on the consequences of the accident, which can be different for each accident. The consequences are measured in terms of deaths of both the public and persons operating the plants, the cost of the accident stemming from the cost of cleanup, replacement of houses and so on replacement power (if appropriate), and payments to the survivors and families of the dead.

The deaths are often grouped into deaths as a direct result of the incident and those that come about sometime later due to the effects of radiation or released chemicals. The deaths in the short term usually result from explosions, fires, and floods, whereas deaths in the long term result from health effects of exposure to chemicals or radiation leading to cancers. For example, the atomic bombs dropped in Japan led to instant deaths as a result of heat and blast effects, very much like conventional explosives, but on a much greater scale. The explosive effects of atomic bombs are measured in megatons of TNT. There were also deaths due to intense radiation effects.

Radiation effects can be due to direct shine effects, but this effect can be ameliorated by shielding provided by walls, houses, and distance. This shielding effect is how nuclear power plants (NPPs) can be safely operated, given the intense radioactivity generated in the central heat sources (nuclear core) used to generate electricity. However, in the case of atomic bombs, radioactivity materials are released into the air. These materials are particles and gases, which can be dispersed over an area and can affect the population, causing different cancers and leading to death, often much later in life.

Nuclear plants can lead to similar consequences if these radioactive materials are released. However, the US industry in particular has developed containment

**TABLE 7.1**
**List of Some Accidents**

| Industry | Accident | Consequence |
|---|---|---|
| *Major Accidents* | | |
| Nuclear industry—USA | Three Mile Island Unit #2, USA, March 1987 | No short-term deaths, at least $2 billion in losses |
| Nuclear industry—Ukraine | Chernobyl, Ukraine, April 1986 | 56 deaths and 2,000 long-term deaths, estimated but not confirmed |
| Nuclear industry—Japan | Fukushima, Japan, March 2011 | 2 deaths at plant, 20,000 civil deaths and 140,000 buildings damaged from tsunami; long-term effects still being estimated |
| Chemical-pesticide—India | Bhopal, India, December 1984 | Some 8,000 died from exposure to methylisocyanate and 500,000 injured |
| Oil and gas industry—drilling operations—USA | BP drilling rig in Gulf of Mexico, *Deepwater Horizon*/ Macondo oil rig blowout massive oil release/fire, September 2010 | 4.9 billion barrels of oil released 5 deaths on the rig, cost of cleanup (billions in fines and other costs), and loss of revenue to persons in the surrounding states |
| Shuttle operations—USA | *Challenger*, January 28, 1986 | Crew killed, shuttle destroyed a few minutes after launch |
| Tenerife, Canary Is., Spain | Pan Am and KLM 747s crash, March 1977 | Planes ran into each other on the ground, 260 passengers and crew died, 5 on the ground, and both planes destroyed |
| *Ancillary Safety-Related Incidents* | | |
| NPP containment sump blockage | Accident occurred at low power, plant safely shutdown | Barsebeck NPP shut down. Many international studies funded to examine safety issue |
| VVER fuel cleaning | A serious accident took place in April 2003 at Paks Unit #2 | Some personnel exposed to radiation, Unit #2 shut down and INS #3 event, cost of unit shutdown |
| Steam generator tube damage on San Onofre units #2 and #3 | Steam break occurred and unit #3 shut down and examined | Large number of tubes on each of units failed NRC criteria, SONGS shut down permanently |
| Northeast Utilities management change, different operational philosophy | NPP start having availability problems, reductions in all personnel to effect O and M costs | NRC ordered all NE plants shutdown, plants sold |

vessels to limit radioactive releases in the case of accidents that lead to core damage. Looking at other industries, one can see that accidents in these industries can lead to fires, explosions, and releases of various chemicals that can also lead to cancers. There are examples of large explosions, such as the incident in Halifax (Canada) in 1917, and even explosions caused by fireworks or stored military arms.

In the case of the release of chemicals, the classic incident was in India at the Bhopal pesticide plant in April 1988, which led to deaths of some 20,000 persons under conditions of great pain due to the release of methylisocyanate (MIC). MIC attacks the soft tissues in humans, such as eyes and lungs, causing very nasty effects.

When one talks about accidents, the public perception is that NPPs are the most dangerous, whereas on closer examination there are other industries that are much more dangerous. The objective here is to discuss decision-making in high-risk industries under stress, not to justify the safety and economics of one industry over another. However, a poorly run company, nominally low risk, can lead to a worse record than a well-run, higher-risk company. It is the quality of the industry's management and staff that can lead to good or poor results.

Some of major accidents are listed in Table 7.1. Whether an accident is major or not is in the eye of the beholder. One measure is the monetary cost; another is the number of persons killed, either directly or indirectly (cancers). Some areas, like nuclear, receive greater scrutiny (exposure in the media) than others. For example, the TMI was the worst nuclear accident in the United States; however, no one was killed and the radioactive impact on the public was very small (small releases of radioactive materials), therefore there was very little direct impact on the public! However, the impact on General Public Utilities was large in terms of loss of an NPP unit, need for replacement power for damaged NPP Unit #2, and for NPP unit #1 being shut down until cause of the accident was understood and the Nuclear Regulatory Commission (NRC) acted to allow its start-up.

List of other interesting accidents/incidents:

1. Space: *Columbia* orbiter accident, November 12, 2002; crew died, leading edge of wing damaged by ice and orbiter burned up on reentry.
2. Air Transport: American Airbus 300, JFK Airport, February 14, 2004; all persons died, rudder/fin failure caused by pilot maneuver to mitigate turbulence induced by Boeing 747 vortices.
3. Petroleum: BP oil refinery, Texas City, March 23, 2005; fire caused by operator action, 14 persons died, $1.5 billion plus cost.
4. Railway: King's Cross, London, Underground fire, November 18, 1987, 23 persons killed.
5. Railway: Metrolink commuter train, Los Angeles, Union Station to Oxnard collided with a freight train in the Chatsworth area, September 12, 2008; 25 persons killed.
6. Nuclear: Davis-Besse near accident due to reactor vessel head corrosive penetration, March, 2002.
7. Hurricane: *Sandy,* Northeast Storm, caused damage to shoreline and flooding emergency power at hospitals leading to evacuation of sick persons.

## 7.4 NUCLEAR INDUSTRY ACCIDENTS

This section covers some of the more severe nuclear accidents: three nuclear cases—TMI Unit #2, Chernobyl, and Fukushima. The approach taken is to have four sections covering the accident: accident description, accident analysis, organizational analysis, and consideration of the viable systems model (VSM) following organization analysis.

### 7.4.1 THREE MILE ISLAND UNIT #2

There are many reports and descriptions for this accident, which was a classic for the US nuclear utility industry, March 1979 (Kemeny, 1979, and Rogovin, 1980). The following accident description is derived from the reports that were read many years ago and have been compared with other authors' reconstructions.

Accident description:

The accident started with a loss of main feed due to an incorrect filter switchover procedure. The TMI NPPs were designed by Babcock and Wilcox and had once-through steam generators (SGs). Attention to water quality is paid for all SGs, but once-through units are particularly sensitive. The main feedwater has in-line filters to improve the feed supply quality, but they need to be replaced at frequent intervals. It was during the switch-over process that feed-flow was cut off.

Both the reactor and the main turbine tripped automatically. In response, the auxiliary feedwater system should have started, but failed to start due to a maintenance error not spotted by the control-room crew, and all auxiliary feed isolation valves were closed. All safety injection (SI) and residual heat removal pumps started due to the correct generation of the SI signal. The reactor primary system heated up, reactor pressure increased, and the pressure-operated relief valves (PORVs) opened. This is the normal response. Subsequentially, the reactor pressure dropped and continued to drop until it reached the saturation temperature pressure and boiling started in the core.

Once the reactor pressure falls below the low pressure set point, the PORVs should have closed. The PORVs did not close, but the operators thought that they had, since the PORV indicators showed them as having closed. Error in indication was caused by poor instrumentation design for the PORV. Boiling in the core continued, and the generated steam rose to the top of the reactor dome and displaced the water there. The displaced water moved into the pressurizer and the level within the pressurizer rose until the pressurizer filled. The operators thought that the reactor pressure was under control and saw that SI was continuing to inject water. The operators thought the change in water level was due to the SI flow, not due to boiling in the core.

Therefore, the operators decided it was not necessary to continue to run the SI pumps and shut them off. The failure of the operators to understand the dynamics of the reactor system confirms Ashby's law of requisite variety. If you are unaware of the variety, you are unable to control the system. In order to cover the core, you need to continue to inject SI flow. The other need was to remove decay heat by using the SGs to cool the reactor. The operators needed to open the atmospheric dump valves to pass the steam so formed to the atmosphere.

The reactor decay heat continued causing boiling, the water continued to drop, and eventually the top of the reactor core was uncovered. As a consequence, the fuel cladding was not being effectively cooled by steam flow, its temperature rose, and clad melting occurred. The clad is one of the three barriers to the release of radioactivity, along with reactor vessel and SG tubing and the containment, meeting the US "defense in depth" philosophy. With the failure of the cladding, some fuel pellets fell to the bottom of the reactor vessel.

Subsequently, the control-room crew, with guidance from a unit #1 supervisor, realized that the core was uncovered and switched on the SI system; unfortunately this further accelerated core damage by shattering the overheated clad, when it was exposed to the cold SI water. The consequence was that the core of the Unit #2 reactor was destroyed and there was a mixture of reactor fuel pellets, and cladding fused together at the bottom of the reactor vessel. The whole unit was written off, a large economic loss, but very few persons were affected, since most of the radioactive material was contained in the reactor and containment.

### 7.4.1.1 Accident Analysis

The operators and the maintenance/testing staff could be declared to be the responsible personnel for the accident. The operators were the "sharp edge" personnel in this case, taking action. However, there were a number of others involved, such as the designers of the plant in specifying these particular PORVs. The PORV was poorly designed in that the signal indicating that the valve was open or closed was derived from the control signal and not the actual position of the valve. So the valve could appear to be closed, but was actually stuck open.

In addition to the operators, others should be held responsible. The industry and NRC leaders should be really held to be responsible in that they did not appreciate how important decay heat was to the safety of NPPs. Both parties had downplayed the role of operators during the accident control and mitigation process. As a result, training of the control room operators was defective.

The management of TMI should also be included in that it was responsible for reactor safety and should have been better trained in reactor, plant technology, and accident dynamics. They, in turn, should have insisted the operators were better informed. However, they represented the norm for the industry. At that time, the best technical knowledge about reactor and plant dynamics resided with the reactor designers. However, they failed to understand the limitations of the operators to understand and control the accident!

Figure 7.1 shows the connections between the various parties and some of the things that went wrong and is a reflection of the previous text. The parties to the accident were the utility and its management, the utility staff (control-room, maintenance, and test), the NRC, and the designer of the TMI units (Babcock and Wilcox). The diagram shows the accident sequence at the bottom of the figure and then relates various actions and decisions taken by various persons, notably by the control-room operators.

The figure shows that the initiating event was the trip of the main feed caused by the filter transfer being performed incorrectly. The closure of the auxiliary feed isolation valves was caused by maintenance staff, which further confused things by

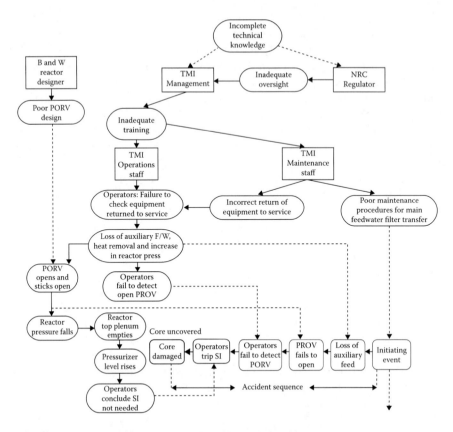

**FIGURE 7.1**    Three Mile Island Unit #2 accident relationships.

incorrect placement of the work tags on the isolation valves. The control-room staff can be faulted for not being in control of the work authorization process for the auxiliary feedwater system tests, which required the isolation valves to be closed only during tests.

It could be said that the control-room staff should have spotted that there was a small loss of coolant going on. However, the indications of the valve position and high temperature indication at the PORV blow-down line exit could be confusing for poorly trained staff, and therefore the blame flows to the utility management and NRC organization. One further thing: the action of the unit #1 operator in analyzing the loss of reactor fluid was negated by him not realizing that the fuel was very hot and that turning on the SI flow, which was cool, could shatter the clad and lead to the fuel pellets falling to the bottom of the reactor vessel. The core damage had already occurred, and the additional fuel and clad fused together at the bottom of the reactor.

### 7.4.1.2   Organizational Analysis

The nuclear industry consists of utilities, manufacture and architect engineers, and the NRC. At the time of TMI #2 accident, there was a lack of understanding of basic accident analysis and the role of control-room operators, especially the role of decay heat removal in accident progression. In the case of the manufacturers, the failure was to transmit their understanding of these issues to the utilities in a clear manner and of the need to pay attention to decay heat removal. It is clear that manufacturers were not aware of the possibility of a lack of competence of operators to control multifailure types of accidents, especially operations involving decay heat. It was thought that the real problem was to ensure that the reactors were shut down. Decay heat was thought to be a plant ancillary state, easily dealt with by the operators, since it represents a small fraction of the heat generated (~10% and reducing).

In the case of accidents involving stuck-open PORVs, a number of operators had dealt successfully with this issue, but information about the issue did not seem to be distributed throughout the industry. Later, one of the functions of the Institute of Nuclear Power Operations (INPO) was to act as a distributor of accident reports and lessons learned, so this was an improvement in the way the industry operated.

At TMI, the management was at fault because training for the control-room operators was insufficient to tackle the TMI-type accident. The maintenance/test personnel were at fault on a number of counts, firstly because of poor processing during changing feedwater filters leading to reactor trip and secondly for not returning the auxiliary feed isolation valves to their open position after performing auxiliary feedwater systems tests. The failure of the main feedwater flow led to the initiation of a reactor and turbine trip. Later, the operators failed to institute heat removal from the reactor via the SGs, because the auxiliary feedwater isolation valves were closed and the auxiliary feed was unavailable. They could have opened the valves and started the auxiliary feedwater system, and this would have saved the situation.

The operators were also at fault for not checking the isolation valves either on shift change or before issuing releases for the "Maintenance and Test (M/T)" work. So there was a management problem here as well. The maintenance and operators' actions were standard relative to returning valves to operational conditions. These actions do not require complex thought processes, just close attention to detail. These actions point to a lack of management attention to operational details together with an unsatisfactory approach to safety.

In addition, the operators seemed to have had a poor understanding of how NPPs behave and a lack of understanding of basic reactor plant dynamics. The operators seemed to not appreciate the fundamentals of heat removal from the core to prevent core damage. The NRC and its personnel were responsible for testing the operators. This process was done frequently. So the NRC was also to blame for a lack of knowledge relative to complex accidents and the role of decay heat in accidents. This lack of knowledge was not uniformly distributed through the industry and seemed to be absent at the key decision levels, like the utility management.

### 7.4.1.3 Review of a VSM Model following TMI Organizational Analysis

A VSM is used to represent organization structures. The structure of the general public utilities (nuclear) (GPUN) organization shown in Figure 7.2 relates to our interpretation of the organization at the time of the accident (March 1987). This diagram is different in some of its details compared with the later NPP organizations, which reflect features that have changed because of the impact of the TMI accident. One

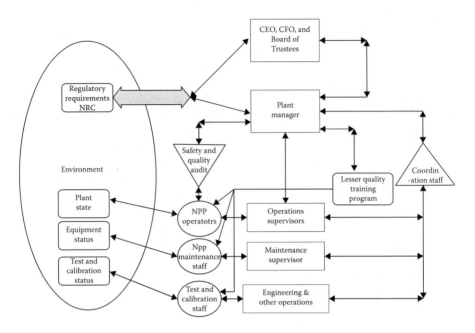

**FIGURE 7.2**     GPUN TMI organization prior to the accident in March 1979 (VSM version).

key item that affected the TMI accident was the knowledge and experience of the TMI operators, and an element has been introduced to represent the greater emphasis given to training, as a marker for this critical function. A director of training is introduced; however, his or her position still ranks below that of plant manager, but including this function explicitly emphasizes the impact of training and simulator training on the safety performance of the plant. It should be emphasized that training for other staff, such as maintenance and radiology personnel, is also important.

The failure of the test personnel to return the auxiliary feedwater isolation valves to their working condition was a failure of the organization: failure to use checking lists (procedures), failure of control by the control-room operators to return to operational conditions after tests, and failure to emphasize safety training. Nowadays, this would be called a failure of the plant's safety culture.

At the time of the TMI accident, the NRC licensed the control-room operators, and the operators had to be trained and tested in responding to various accident sequences. Mainly this training was carried out in a classroom, since there were a small number of simulators available. This clearly indicated that simulators were not universally considered a necessary element in the training of control-room operators.

The role of simulation in the manufacturer's design groups was considered totally necessary for designing control and protection systems and studying accident progression on the safety of the plants. It was through the use of simulation that the design groups could ensure that Ashby's law was satisfied, although Ashby's law was not explicitly known at the time by designers. However, they were aware of the need to represent the plants dynamics as close as possible, so they turned to mathematical simulation of the plant dynamics. It was as though the designers knew of the need to satisfy Ashby's law implicitly.

The situation was such at the time that the control-room operators might even be trained on simulators that were not duplicates of their plant. During the TMI accident period, there was a limited number of plant-specific simulators and they were not full scope, in that the secondary side (feedwater/steam) was not dynamically modeled. Also, in the early days of simulator training and testing, design-basis accidents were selected for training, as opposed to the later approach of exposing crews to multifailure transients. The earlier approach used to prepare the operators was approved by the NRC, and it was founded on the belief that the automatic protective functions would protect the plant and the public and that the role of the operators was less critical. In retrospect, this was thought to be wrong.

Following the TMI #2 accident, the industry made a number of changes in the NPP organizations, and these are shown in Figure 7.3.

There are both visible and hidden differences between Figures 7.2 and 7.3. The two most visible differences are the presence of INPO and the management position denoted by the CNO, or chief nuclear officer. If one studies the two VSM models, one will realize that many of the functions indicated by the same name are in fact performed differently. The existence of INPO is directly due to the effect of the TMI accident on the industry (Rees, 1994). But the TMI accident's influence was more invasive than just the development of INPO. The whole position of humans within the industry changed as the messages stemming from reports generated in response to the accident; namely Kemeny (1989) and Rogovin (1980) became known.

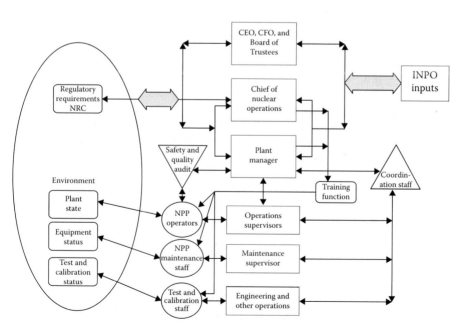

**FIGURE 7.3** Depiction of a VSM model of a US nuclear utility.

In particular, there were many things aimed at improving the performance of the control-room operators. The following lists some of these:

1. Improvement in "Emergency Operating Procedures" from event-based to symptom-based.
2. Requirement for each station to have a full-scope simulator for each different reactor type at the station.
3. Each control room should be reviewed for human factors' compatibility with the needs of the operators to better respond to accidents.
4. A person trained in engineering to assist the operators in the analysis of symptoms following an accident. The person was a "shift technical advisor."
5. A display tool, called safety parameter display system (SPDS), to help the operators define which parameters should be key to their understanding of an accident, as it progressed.

Requirements have been added for all persons working in NPPs, in response to the actions taken after the TMI accident. All parties making decisions need to be well founded in knowledge of plant behavior and need to be well trained in the nuclear arts. In the case of the utility management, there is a need to be aware of NRC regulations and an understanding of fundamental reactor safety, including such things as "defense in depth." The top managers are responsible for running the plant, selecting good personnel, and getting the plant to run efficiently (cost control) and safely. Well-trained and competent personnel are required throughout the NPP organization. The top managers need to be involved in all aspects of the plant operations, and this seems to be a key need to ensure plant and radiation safety and economic operation.

Top managers do not seem to be held to same high standards of NPP safety as are the control-room operators, but it appears that managerial staff are now more aware of nuclear safety requirements than before. INPO seems to have promoted the idea for top managers to get training in NPP safety. Adm. Rickover was very careful in both picking and training officers to serve on his submarines. He was ahead of others in the process of preparing his officers for increasing responsibility. Adm. Rickover's approach has had an influence on INPO, since many of managers at INPO have come from the US nuclear Navy.

In fact, most if not all high-risk organizational (HRO) industries seem to be lagging in the process of preparing company officers by exposing them, during training sessions, to increasingly difficult decision-making situations. One is not born with the capability of good decision-making; this is a learned skill and does not come with the job.

### 7.4.2 CHERNOBYL

The Chernobyl accident took place at Pripet, Ukraine, on April 28, 1986. The accident led to the complete destruction of the Unit #4 reactor. The reactor was of the Reactor Bolsho-Moshchnosty Kanalny (RBMK) design. In Russian, RBMK means large multi-channel reactor. A different approach is going to be taken for this nuclear accident compared with TMI #2 and Fukushima descriptions.

### 7.4.2.1 Description of Plant

The reactor design is a graphite moderated reactor with cladded fuel cooled by water. Figure 7.4 shows a schematic illustrating the plant configuration. In some ways, it is similar to the early British and French Magnox reactors, except that the Magnox reactors were cooled by carbon dioxide. All of these reactors were physically large. Unlike the US-designed PWRs, none of these reactors had an enveloping containment, but had some reactor partial containments. The RMBKs have limited containments around the hot leg coolant piping; these were completely inadequate for the actual accident sequence. The plant has two steam turbines, multiple headers dealing with both water feed into the reactor and mixed steam/water flow out of the reactor. The flow out of the reactor goes to a number of drums in which the steam flow is separated from the water. Water in the drum passes via reactor pumps back into the reactor. The water level in each of the drums is made up from cold feed from the turbine condensers. The diagram shows a main feed pump, there are a number of such pumps.

RMBKs were designed as both power and plutonium producers (for bombs). The plant was refueled online, so that the fuel could be used optimally for reactivity control and plutonium production. Part of the design was to use control rods with graphite extensions. This design feature was to help combat the fact that the reactor had a positive void coefficient. In fact, the USNRC does not allow reactors to be built with positive coefficients. However, the reactivity coefficient for the fuel was negative. Positive reactivity coefficients can lead to runaway power escalations, and this is what occurred in this accident.

### 7.4.2.2 Accident Description

The accident started in a most innocent manner with a test that had been carried out before, but because of delays in arriving at the required test conditions, the operator trained to carry out the test was replaced due to a shift change. The test was to see

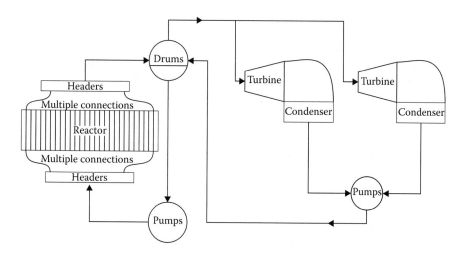

**FIGURE 7.4** Schematic of the Chernobyl (RMBK) reactor.

if the delay associated with the diesels coming online could be compensated for by using the main turbine generators to continue to power the reactor circulation pumps. In the previous test, which was unsuccessful, there was a problem with voltage control and the voltage regulator design was changed in response to the failure.

However, the experiment was delayed by the load dispatcher because of a need for power. This led to the shift change that meant the person most knowledgeable about the test requirements was not available. So this was the start of the problems. Also, RBMKs are difficult to control, even unstable under some operating domains, because of the positive void coefficient, which means that if voids (steam) are produced, the power increases, leading to more voids.

Quite clearly from the number of commentaries, the consensus is that the operator was to blame for the accident. Of course, it is easy to blame a person who is no longer able to defend himself. We have tried to construct a sequence of events from the descriptions available in various documents. In general, the main focus of many documents has been the consequences of the accident, that is, the number of people who died (28–34) and those who might die of cancer; estimates vary in the thousands. There was contamination of the soil around the station and beyond, even to the grass in the Welsh hills (UK). It is interesting to note that the Russian (Soviet) government said very little at first and the Swedes (high levels of radiation detected at Forsmark NPP) were the first to detect the radiation fallout. It is believed that the operator did try to follow what was requested of him, but the operators were creatures of the training processes used under the Soviet government.

One could say that the Chernobyl accident was more the result of the design processes and decisions taken in terms of operating the plant than it was just human error, although the actions were taken by the operators. The accident was one that was waiting to happen: if not Chernobyl, then some other RBMK plant. It should be noted that even for the other Russian-designed reactors, the VVERs, the latest versions have containments. Also the training methods being used for the surviving Russian-designed plants have also moved to be in conformance with Western practices!

The test was started with the thermal power being changed to 700 MW from 3200 MW, since the test was to be carried out at a lower power. The crew reduced power too quickly and there was a mismatch in the xenon 135 production and decay. Because of the crew's relative lack of knowledge, they did not know of the relationship between power and xenon poisoning, and the crew continued to carry out test operations.

### 7.4.2.3   Accident and Organizational Analysis

On April 26, 1986, at 01:23, Reactor unit #4 suffered a catastrophic power increase, leading to explosions in its core. The head of the reactor vessel was lifted and this dispersed large quantities of radioactive fuel and core materials into the atmosphere and ignited the graphite moderator. The burning graphite moderator increased the emission of radioactive particles, carried by the hot gases into the atmosphere. Unlike Western reactors, there was no containment to prevent the release of radioactive material. The accident occurred during the experiment scheduled to test the ability to use the power during shutdown to provide power to drive pumps.

The information related to the accident and the effects of the plant organization on the accident are somewhat limited, and it is not as to analyze the different effects as it was for TMI #2 and the Fukushima accident.

However, the following points could be made from information released through one source or another:

1. The experiment design seems to have been poorly considered from a risk point of view.
2. The replacement crew was not trained in the processes needed for the test. From a prudent viewpoint, the tests should have been delayed until the more experienced operator was available. It could be that test still would not be safe, because of the stability issues associated with the core and rod design.
3. There did not appear to be a clear understanding of the dynamics of the core as far as xenon transients and thermal effects were concerned. This points to a need to understand Ashby's law as applied to this system, in order to better control the reactor under these conditions.
4. The crew bypassed safety functions in order to meet the needs of the experiment. This was a clear violation of fundamental plant safety.
5. The plant conditions changed as a direct result of the delay caused by grid distribution requirements. The effect on the plant conditions, because of the delay, did not seem to be factored into the planning of the experiment.
6. Plant management did not appear to be in charge of the plant to ensure the safety of plant personnel and the public. It did not appear that they were involved in the details of the planning and execution of the experiment. If the test was delayed, they should have determined that the replacement staff was not adequate for such a test, despite the fact that an earlier test had been carried out (unsuccessfully)!

### 7.4.2.4 Comments on Chernobyl Organization

In the TMI #2 discussion, we included VSM diagrams to discuss the characteristics of the GPUN organization at the time of the accident. We could make an attempt to derive the Chernobyl NPP organizational functions. It appears that not much is to be gained by doing this, since it seems that the basic experiment was determined by a Soviet central organization trying to carry out the test: the NPP management personnel were directed to carry out the task and the operator was selected to perform it. Unfortunately, the need for power overrode the test and affected the conditions. So decision-making was carried out by the central authority, and the local management was ordered to proceed with the test!

Once the accident took place, the core was quickly reduced to a burning mass and the Soviet government acted to prevent the continuing release of radioactivity by dropping a mixture of cement and lead pellets, presumably to try to prevent the melted mess of fuel pellets from returning to critical. The helicopter crews performing this duty suffered from radiation effects.

Looking at the balance between the various organizations involved, it seems that there were conflicts between these various bodies: the plant management, the research group, and the grid control personnel. There should have been an agreed plan to run the tests in a manner to ensure plant safety. This does not seem to have had any importance in the minds of any of the groups. This lack of understanding, and allowing the initial conditions for the tests to be overridden by the grid requirements, is what led to the accident! One presumes the underlying failure was the failure to point out the concerns related to the stability of RMBKs, as far as it affects what exercises can be carried out safely. All of the persons involved seemed unaware of this safety issue, or perhaps the advice of some was not taken. This is another situation for the invocation of Ashby's law. They were trying to carry out tests without understanding the variety of the system under off-normal conditions.

### 7.4.3 FUKUSHIMA DAIICHI ACCIDENT

The accident referred to as the Fukushima accident took place in Japan on March 11, 2011, and affected a number of nuclear plants operated by TEPCO. The plants were the six units of the Daiichi and Daini stations and are about 160 miles north of Tokyo on the northeast coast. Four of the six plants that made up the Daiichi were the ones principally affected. The accident was caused by a series of large earthquakes and later followed by a number of large tsunamis. The largest earthquake and the some of the tsunamis exceeded the design bases for the NPPs.

The Fukushima accident has had a large effect on the nuclear community. The responses taken by the crew to the accident modified the normal organizational structure to one reflecting the needs of the site to combat the effects of the accident. Before discussion of organizational impacts, details about the accident need to be addressed. Analysis of the accident will be carried out and lessons as far as organizations will be covered. Two sets of reactors owned by TEPCO on the northeast coast of Japan were involved; these were Daiichi and Daini, Dai first and second stations. The reactors were initially affected by a large earthquake (9.0 on the Richter scale) followed about one hour later by devastating tsunamis. The earthquake appears to have done some damage, but the biggest effect on the reactor units was caused by the tsunamis. A short description of the accident sequence is given, but the emphasis here is on the roles of the plant operators, the plant management, TEPCO management, and the Japanese government as well as can be estimated and then related to organizational models (VSM) for diagnostic purposes.

### 7.4.3.1 TEPCO and Fukushima Plant Organizations

The three organizations closely associated with the Fukushima accident were the Japanese government, Japan Nuclear Energy Safety Organization (JNES), and the Ministry of Education, Culture, Sports, and Technology (MEXT). Following the accident, there have been changes in the Japanese organizations in response to postaccident analyses of perceived deficiencies in the organizations' response to the accident.

TEPCO organization shifted the focus of the management site organization responsibly related to Fukushima nuclear plants from operation to clean-up of the site. The units are unlikely to operate ever again, because of the extensive damage and long-term effects of radiation release from the destroyed cores.

The site organization for the Daiichi units is shown in Figure 7.5. This figure was retrieved from the INPO report on the Fukushima accident report (INPO, 2011). The titles and working positions are not identical to those of a US NPP organization (see Figure 7.3). The site superintendent appears to be equivalent to a US site vice president, the unit superintendent is equivalent to a plant manager, the operations general manager to an operations manager, the shift supervisor to a control-room supervisor, unit senior operator to a reactor or at-the-controls operator, unit main shift operator to a balance-of-plant (BOP) operator, assistant senior operator to a field supervisor, and auxiliary operator to a nonlicensed operator.

### 7.4.3.2 Comments on the Preaccident Status

Before one can totally understand what happened, one needs to understand some of the underlying history with respect to TEPCO and a little about the reactor designs. It has been recognized by various parties, such as the Japanese press and even organizations such as World Association of Nuclear Operators (WANO), that TEPCO management was not the best in the industry. The boiling water reactors (BWRs) at the Fukushima site were relatively old, being designed before the 1970s, and several upgrades had been recommended to deal with a number of issues, including

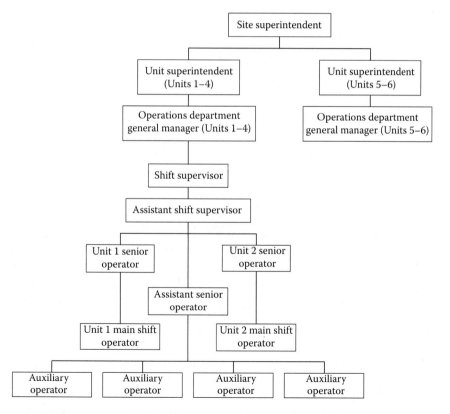

**FIGURE 7.5** Fukushima Daiichi NPP organization. *Note:* Units 1–4 each have common control rooms.

hardening vents to avoid the possibility of hydrogen explosions that could cause the failure of the reactor building. In addition, the Japanese press reported that TEPCO's operations were not good. It was even suggested that certain tests of containment leakage rates were carried out incorrectly.

All NPP designs have been updated to ensure that they are better able to meet the latest safety standards. People have also reviewed the bases for external events to see if more care is needed to ensure that NPPs can safely ride through all external events, such as earthquakes, floods, etc. Studies by organizations such as the US Geological Survey have pointed out new fault lines that have been discovered and these new faults may lead to a higher ground acceleration condition than that the plant was designed for—the so-called design-basis event. The regulatory authority then orders that changes should be made to meet these new situations.

In the case of TEPCO, it was warned by Yukinobu Okamura (head of the Active Fault and Earthquake Center), some 2 years before the Fukushima accident, that the site could be threatened by a tsunami greater than the design-basis event (CNN, 2011). TEPCO has been accused of not being very open to questions (Reckard, 2011; Shirouzu and Smith, 2011) and was found to have falsified records some time ago. TEPCO appeared not to take the suggestion about the tsunami very seriously and did not increase the seawall height or waterproof the NPP electrical installations. The managerial elements involved in the process of assessing and making safety changes are TEPCO management, the Japanese regulator (NISA), and the Japanese government represented by the METI. The government organizations did not push for changes to be made!

### 7.4.3.3  Accident Description

The large seismic event (Richter scale 9.0) that occurred in March 2011 off the northeast coast of Japan was above the design bases for the plants and caused a large amount of damage, including affecting electric power distribution, and led to the automatic shutdown of the Fukushima NPPs (Daiichi). This was an entirely accepted response. The actual ground acceleration was 0.56 g versus design-basis 0.447 g. The standby diesels started up and the plants were operating safely. Of the six NPPs of Daiichi, only units #1, #2, and #3 were operating; the other three NPPs were shut down for various reasons and were not operating. The seismic analyses tend to be basically conservative, so plants can take ground acceleration higher than the design-basis number.

As the result of the type of earthquake off the coast (a subduction fault), a series of large tsunamis was generated (INPO, 2011). The INPO report is very detailed but still does not address questions related to why certain actions were taken. The tsunami caused a lot of devastation to the area around the region where the NPPs were located. Many people were killed and their property was destroyed. Roads were swept away and rail transport ceased, along with a loss of communications.

The INPO report indicates that there were multiple tsunamis, some seven altogether. It also states that several aftershocks of lower magnitude occurred before the tsunamis arrived. At least one of the waves was approximately 46–49 ft (14–15 m), based on water level indications on the buildings. The design-basis tsunami was

18.7 ft (5.7 m) above ground level, so the largest tsunami was well above the design-basis. Figure 7.6 shows the various measurements related to the building and water levels reached during the tsunami. In addition to the previously mentioned earthquake damage, the tsunamis were of such a size that they overflowed the NPP sea-wall protection, which was supposed to be bigger than the design-basis for the NPPs.

The magnitudes of both the earthquake and tsunami exceeded the design bases for the NPPs. It has been reported that seismic experts had informed TEPCO about 2 years earlier that this same area was devastated in AD 875, by a tsunami of a similar size to this one and that information should have been included in the database from which the design-basis tsunami was selected (CNN, March 27, 2011).

Because of the size of the tsunami, seawater caused the standby power diesels to fail, the diesel fuel tanks to be blown away, and some battery rooms and levels in turbine halls to be flooded (see Figure 7.7). There were some diesels that were air started, but could not be used, since the rest of the electrical systems had failed. The grade level for reactor buildings was at 33 feet, but the electrical equipment, switch gear, batteries, and emergency diesel generators were below grade level. The inlet cooling system structures were at 13 feet and they became blocked with the debris caused by the tsunamis, which led to cooling water pump failures.

The loss of diesels and battery supplies led to the plant being put into a "blackout" condition. A blackout is a situation in which both offsite power and station-generated power are lost. The station-generated power is derived from standby diesels, which are supposed to start up on the loss of offsite power within a short time. The usual mechanism considered for the loss of diesels is a failure of the diesels to start related to diesel generator failure mechanisms. In this case, the diesels started and then stopped due to the tsunami flooding the diesel locations.

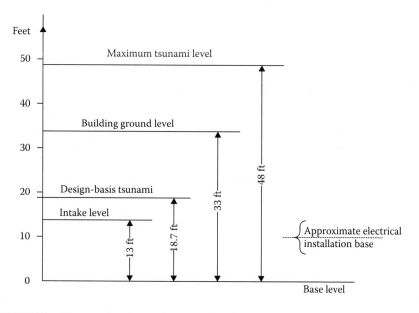

**FIGURE 7.6** Diagram showing various water levels.

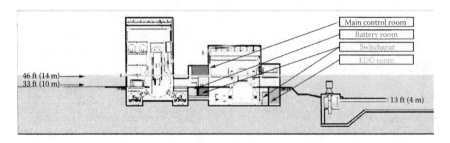

**FIGURE 7.7** Diagram of NPP showing general elevation and flooding level during tsunamis.

The reactor control-room personnel were alerted to an emergency condition, with four units in a hazardous condition. Even well-trained operators, with a well-developed emergency plan, would have a great difficulty in knowing what to do and they had very little time to take action.

Initially, all went well following the earthquake, the reactors shut down (control rods inserted into the reactor core), the auxiliary electric supplies via the diesels came on, and the initial stages of decay heat removal were taken care of.

There may have been some damage from the earthquake: but it did not lead to extensive damage at the plant. However, within an hour of the earthquake, the tsunami struck and from then onward, the safety systems failed and the batteries failed to supply instrumental power to allow valves to be operated. Under these conditions, it was nearly impossible to prevent core damage and loss of cooling to the spent fuel pools.

The crews' action was then to try to reduce the pressure in the reactors to a point where they could use the fire pumps to inject water (initially freshwater then seawater) into the core. The crews were also faced with the fact that their families and friends might have been killed by the effects of the earthquake and the tsunami. The site superintendent was involved in the stabilization process, but it appears that the emergency procedures that were practiced were not designed for such difficulties. Confusion abounded in and around the plant, and resources to help the personnel were not readily available.

In the surrounding areas, people were killed and injured, houses were damaged, transportation affected, cars washed out to sea, etc. It is believed that somewhere in excess of 20,000 people died and more than 140,000 houses were destroyed (Japan Fire Department, 2011).

A number of accident reports relating to Fukushima were available shortly after the accident (for example, see Braun, 2011 [AREVA]), but they focused on the accident progression, what actions were taken, and what the state of the plants was at various intervals. The reports were classical in that they focused on the accident sequence—giving information about what was going on, such as hydrogen explosions occurring in the region of the various spent fuel pools, radiation releases, etc.

Very rarely does one get a glimpse of what was going on as far as instructions to operators from plant management, TEPCO upper management, the Japan government, etc. Of course, instructions might have had little effect initially, in that the plant was already in a state where the operators could not determine what actions could be taken, since there was no electric power and battery power to instruments and controls also quickly disappeared. Truly, not only was the plant in a "blackout," but so was the operational staff.

TEPCO's top management seemed to be out of touch during the early stages of the accident. It is presumed that advice and help were slow in arriving. The Japanese government was very involved in trying to establish control over the affected regions. The figures are that some 20,000 people were killed, many more were injured and missing, and large tracts of houses were destroyed. It was a huge catastrophic event for the people of Japan.

It is no wonder that even the issue of a reactor disaster was not immediately given enough attention and resources to terminate the accident and mitigate the effects of core damage. In some ways, the site personnel did very well to stay and try to address the problems. It is not clear that the NPP staff and managers recognized the possibility that given the failure of spent fuel cooling, the water covering the fuel would boil away and the fuel cladding would heat up and react with the steam and form hydrogen. Photos of the reactor buildings indicate that hydrogen explosions had taken place. Later, ground personnel were seen pumping water into the direction of the spent fuel pools, which are high up in the remains of the reactor buildings.

The general impression is that the local NPP personnel were overwhelmed by the events but were trying as best as possible to cope with the situation. TEPCO headquarters personnel could not help to improve the situation. Subsequently, radioactivity spread throughout the area. Some of it was airborne and some spread through leakage from the reactor building and spent fuel pools. The full story is not available as to where all of the sources of radioactivity were located. It is believed that some parts of the reactor vessel and its containment system were impacted by the earthquake and a leakage path to the sea could have come from there, as well as from other sources. It might be some time before a complete accounting of the accident sequence and the sources of radioactive releases are agreed upon.

The INPO report covers some of the difficulties that the site personnel had. Included here are a couple of paragraphs to give an insight into the problems that the personnel had. The locations were dark, radiation was high in some locations, equipment was not working, earthquakes caused vibrations, and the threat of explosions existed. This extract is from INPO report dealing with Unit 3:

> *The operators understood they needed to depressurize the reactor but had no method of opening an SRV. All of the available batteries had already been used, so workers were sent to scavenge batteries from cars and bring them to the control room in an attempt to open an SRV.*
>
> *At 0450 (T plus 38.1 hours), workers attempted to open the large air-operated suppression chamber containment vent valve (AO-205). To open the valve, workers used the small generator to provide power to the valve solenoid. An operator checked the valve indication locally in the torus room, but the valve indicated closed. The torus room was very hot because of the previous use of RCIC, HPCI, and SRVs; and the room was completely dark, which made a difficult working environment. By 0500, reactor pressure had exceeded 1070 psig (7.38 MPa gauge), reactor water level indicated 79 inches (2000 mm) below TAF and lowering, and containment pressure indicated 52.2 psia (0.36 MPa abs).*

Later

*A large hydrogen explosion occurred in the Unit 3 reactor building at 11:01 on March 14. The explosion destroyed the secondary containment and injured 11 workers. The large amount of flying debris from the explosion damaged multiple portable generators and the temporary power supply cables. Damage to the fire engines and hoses from the debris resulted in a loss of seawater injection. Debris on the ground near the unit was extremely radioactive, preventing further use of the main condenser backwash valve pit as a source of water. With the exception of the control room operators, all work stopped and workers evacuated to the Emergency Response Center for accountability.*

*The acronyms are: SRV =Safety Relieve Valve, Torus is part of the containment of a BWR, RCIC = Reactor Core Isolation Cooling and HPCI = High Pressure Coolant Injection, TAF = Top of Active Fuel.*

Given that a blackout had occurred and that the tsunami had impacted the site (with roads made impassable with debris and even oil tanks moved by the force of the tsunami), the station staff tried very hard against odds to cool the reactors and cover the reactor cores. The loss of power affected not only pumps and valves, but lighting and availability of instrumentation. For example, the staff did not know the water level in the reactors. The crews located car batteries and connected instruments to determine reactor water level. As a side issue, it is considered that this information was erroneous due to voiding in the reference legs of the level instruments.

The site personnel were faced with the fact that given almost nothing worked, the question was what pieces of equipment could be placed into some degree of working order and what actions did one have to take to accomplish this? This is carrying out emergency planning on the fly, as one can see from the second paragraph above.

### 7.4.3.4   TEPCO Daiichi Organization—Prior to Accident

Prior to the accident, a VSM model of the TEPCO organizations associated with the Daiichi and Daini Stations would look similar to VSM plant models for NPP organizations. Figure 7.8 shows the organization from the TEPCO chairman to the staff of Daiichi #3 unit.

The CNO role in US NPPs is important in that it brings emphasis to nuclear safety and a better balance between economics and safety. If the TEPCO's top management had looked at the risk of a tsunami over the design-basis size, they would have not have failed to take action to increase the size of seawalls, move diesels to higher ground, and have better water protection for their electrical equipment.

INPO is a US organization, but the WANO functions somewhat like INPO and TEPCO is associated with WANO. It appears that the Japanese utilities do not feel that the relationship with WANO is sufficient and have stated that they will establish a new organization under the auspices of the Federation of Electric Companies of Japan (FEPC) The chairman said, "We intend to create an environment that proactively accepts evaluations and advice from external perspectives so that the new system will function continuously in an effective manner rather than becoming a mere façade", (FERC, 2012). This implies that in the mind of the Japanese utilities, the net job done by WANO and TEPCO was not satisfactory: *"a mere façade."*

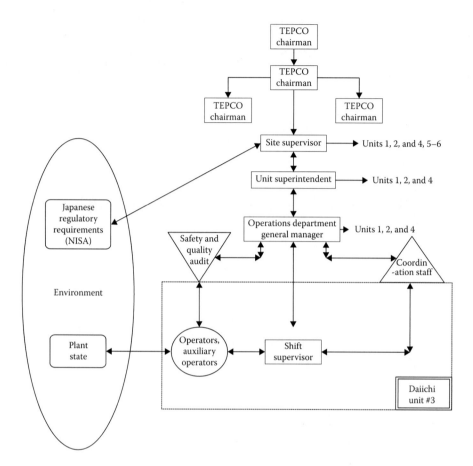

**FIGURE 7.8** VSM model of the TEPCO site organization for Daiichi Unit #3 prior to the accident.

It is suspected that the issue was with TEPCO, but maybe WANO was not persistent enough to impact TEPCO. Not having a CNO position means that safety issues are not as strongly considered as they ought to be. Following the accident, the organization dealt with the reactor trip and loss of off-site power. The Major Disaster Management Headquarters was established after the accident. Initially, the response was carried by the control room operators. Warnings relative to the situation were issued by the site management to TEPCO management.

### 7.4.3.5 Reorganized Daiichi during Response to Emergency

Soon after the tsunamis hit the stations, electric power was lost, leading to a station blackout, and the staff morphed into one dealing solely with responding to the consequences of the worsening accident. The Emergency Response Center (ERC) was set up and the emergency plan was entered based on loss of AC power. Coordination of the responses for all Daiichi units was via the ERC with the site superintendent directing operations, including investigations into how to inject water into the reactor to cover

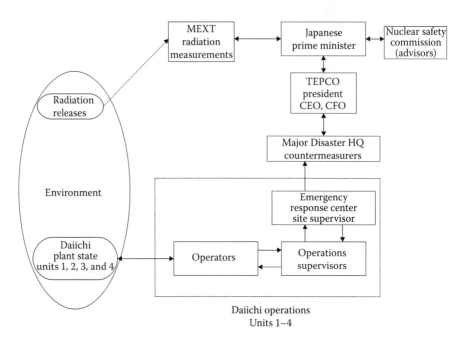

**FIGURE 7.9**    VSM model of the Daiichi emergency operation organization.

the core. Operators at each of the units were trying to energize various instruments and open valves in response to the need to operate vents, depressurize the reactor, try to start pumps, isolation condenser, etc., depending on the assessed need of the specific unit and as directed by the ERC. So the VSM structure corresponding to this situation differs from the VSM for the normal operation. The VSM model of the Fukushima Daiichi response structure to the accident is depicted in Figure 7.9.

This structure was much tighter than the normal organizational structure, commensurate with the need to make local decisions depending on the state of the plant and equipment. The bulk of the decisions and actions were taken at the lower levels of the organization. However, in one case the crews waited to take an action, not wishing to place the local population at risk if the reactor's containment was vented. Most of the public had been evacuated, but some people had not left. If the population was there during a release, it is likely that that they could have been exposed to radiation effects. The crews requested permission to open the containment vents to release hydrogen, and the Japanese prime minister gave permission.

## 7.5    CHEMICAL INDUSTRIES

### 7.5.1    Union Carbide Sevin (Pesticide) Plant, Bhopal, India, 1984

#### 7.5.1.1    Introduction

The accident at the Bhopal pesticide plant was the worst chemical plant accident in the world. The plant is located outside of Bhopal, which lies in the center of India more or less on a line through Kolkata (Calcutta) and much to the north of

Mumbai (Bombay). The plant was constructed by Union Carbide (UCC), owned by the Indian government, and run by Union Carbide India (UCI). The plant was designed to produce Sevin pesticide. A material used in the production process was methylisocyanate (MIC). It was a reaction between water and MIC in a tank that led to the accident.

It should be said that the market for Sevin was not as prosperous as it once was and that production was down, which led to economies in staff and operations. The whole circumstance of the accident and its precursors is very obscure. Some attempt is given here to point out some of the precursors during the discussion about the accident itself. Given the severity of the accident, one would have expected that everyone concerned would have been anxious to try to find out what actually happened.

The Indian government prevented witnesses from being questioned until some-time after the accident and access to documents was prevented. All of this went on for a year, until the US courts forced the issue. By this time, the witnesses were dispersed and the memories of the various participants, if located, were different. Also, records had been altered to protect—whom? The measures taken have affected one's capability to understand the accident and have significantly affected the possibility that one will ever know much of the details. Accident investigations are difficult at the best of times, but the delays, distortions, alterations, and loss of memories make this case not only the worst, but also the one most difficult to sort out. The various reports about the accident are somewhat uniform in some of the facts: that an accident occurred, that it was caused by the explosive release of MIC, that the plant was poorly run, and that a lot of people got killed in a very nasty way. Even the numbers of people that died shortly after the accident and later are ill defined.

One can be sure that both numbers are large and some 2,000–200,000 have died or been injured. Many reports have focused upon the effects of MIC on people and animals. It is not intended to discuss this here, beyond the statement that the effects are very bad and that MIC attacks the soft tissues, including lungs and eyes, and causes great pain. Also, it appeared that the authorities allowed people to move into what should have been an exclusion zone, thus making things worse. It may not have prevented that many being affected, but maybe the numbers of people affected could have been reduced to 100,000 or less.

### 7.5.1.2 Accident Analysis

The cause of the accident was that a significant amount of water entered into one of three tanks containing some 50 tons of MIC. During normal operation, the MIC would be cooled by a cooling system. At the time of the accident this system was not working due to the unavailability of the coolant for the refrigeration system and this may have made the reaction more explosive. How the water got into the tank is subject to some uncertainty. One concept was that the water entered during a cleaning of pipes, and another was that water was deliberately poured into the tank from a nearby hose via the connection from the removal of a pressure gauge.

The latter suggestion has some credibility, since the working relations between the management and the staff was poor, due to the uncertainty that the plant would continue to be operated, given the market for pesticide. According to some reports,

UCC was thinking of moving the plant to either Brazil or Indonesia. In this atmosphere of uncertainty, one worker who was having difficulties with his manager became the candidate suspected of sabotage.

Since the independent in-depth investigation was late in being undertaken, it is difficult to prove this one way or another, unless someone confesses. A logical discussion about one way water entered into the tank can be dismissed, because of issues associated with the available pressure head from a hose used to wash down some distant pipes. Also, an examination of a pipe leading to the tank was found to be dry, indicating the water did not come from that direction.

The most believable act was that someone did connect a hose close to the tank in order to cause problems. Further, it is thought that the person(s) did not really appreciate the extent of the damage that the accident would cause.

The first indication of an accident was a release of MIC. The action of the plant personnel was to try a number of different things to mitigate the accident, most of which did not work. The state of the plant was poor due to inadequate maintenance, and lack of materials hampered recovery operations. The staff tried to transfer MIC to other tanks, spray water to minimize MIC releases, and initiate alarms to encourage the nearby population to disperse. They tried to mitigate the effects of the accident, but in the end they left in order to escape the effects of MIC. The plant was badly designed, poorly run, and there were inadequate safety precautions, such as better containment of dangerous materials, inadequate sprays, dump transfer capacity, and poor warning and evacuation procedures for the nearby public. It is interesting that UCC and UCI did not have sufficient money to run the plant properly, but were able later to pay out $470 million (court settlement) plus $118.8 million in other monies. This should be an indication to management to pay attention to operational practices and design details! This is our point all along. Management has to be trained and tested to understand the impact of accidents on both people and the economics of running a plant. The cost of shutting the plant down safely is probably a whole lot less than paying damages. It appears that nothing can truly compensate for the pain and agony suffered by the Indian people around the plant!

One cultural issue that remains to be discussed is the almost universal engineering culture about the view of designs. Engineers seem to design things and assume that their designs are going to work properly. They also need to think of what will happen if they do not! Application of hazard and operability study (HAZOP) can be very useful for "what if" situations. In this case, if the combination of water and MIC had been thought of in this manner then several design features would have been added. Even the idea that the only way for water to get into the tank was by sabotage and every other way was impossible indicates not only the mind-set of the design and operational personnel, but also of the investigators. The protection systems should have been working on the basis that things can go wrong, especially with substances like MIC. The words of Lord Cullen (Cullen, 1990) mentioned in the case of the Piper Alpha accident come to mind relative to the use of nuclear plant design considerations as far as the safety of the plant is concerned.

A comment on the fact that there were a number of features that should have helped the personnel prevent or mitigate the effect of introducing water into the

MIC tank: The problem is that the defense methods did not work properly or were not available; therefore the result would have been the same. The other fact was that operator actions did little to affect the accident progression and its consequences.

The net effect was that the operator responses were not successful in the prevention of the accident or the resulting deaths. The prevention of the escape of dangerous chemicals needed a complete change in the design and operational aspects of the chemical plant. Also, it might need the chemical industry to be regulated more like the nuclear industry and to have a trade organization the equivalent of INPO. The Bhopal plant needed a number of improvements in design and operation in order to prevent the release of MIC. It appears from the descriptions of the actions taken by the staff that they did their best to try to minimize the effects of the accident. From the reports, they is not clear that the alarms worked but the staff did respond to the effects of MIC. They tried to transfer MIC to another tank and tried to use sprays to control the release or reduce the extent of the release.

The warning for the city dwellers was sounded. Unfortunately, the warning was terminated by order to prevent the panic of the townspeople. So it was better to let them die rather than warn them! Very strange logic! It would appear that beyond the personnel problems at the plant, the staff did take actions to try to control the effects of the incident.

The message that one gets from the review of the accident is the irresponsibility of the management not to maintain the plant in good working condition. If it did not want to pay to operate the plant efficiently, then the decision should have been made to shut the plant down. The question about Indian government politics enters into that decision, and this may have been the reason for not shutting down the plant. It does seem to be very suspicious that the government controlled all releases of information after the accident for at least a year.

### 7.5.1.3 Organizational Analysis

1. The central management was in the United States at a distance from the plant. There appeared to be control issues between the Indian government and US management.
2. Local management was in a difficult situation and seemed not to be in control of the situation as a result of management/personnel relations. It appeared that management/personnel relations were not very good at the plant.
3. Management was not very responsible toward running an efficient plant by ensuring the plant was well maintained, such as ensuring safety systems worked, personnel was provided with good training and procedures, evacuation alarms worked, and the processes were practiced and involved the nearby Bhopal citizens.
4. Despite the problems at the plant, the staff tried its best within the limitations of the plant and its systems.
5. It does appear that the most likely cause of the accident was the disgruntlement of the staff, and of one person in particular. It is likely that he did not appreciate the extent of his action.

6. Given the plant condition, plant manning should have been more carefully controlled so that one miscreant was not in a position to cause so much harm. There is enough history to know that sabotage can occur under poor industrial situations. Management needs to ensure that if it does occur, it is only a nuisance and does not result in a catastrophe.
7. The management had been warned during previous visits of the poor safety configuration of the plant. For example, the MIC tanks contained large quantities of MIC and it would have been better to have had a larger number of small-capacity tanks.
8. If a good safety analysis of the plant had been carried out, it might have helped, but that is unlikely to have occurred given that the management was basically irresponsible in its attitude toward safety and health of the public.

## 7.6   OIL AND GAS INDUSTRIES

### 7.6.1   *Deepwater Horizon*/Macondo Blowout Gulf of Mexico Oil Accident

#### 7.6.1.1   Introduction

This accident is included because it involves some interesting decision-making, not only by British Petroleum (BP) personnel, but also in the participation of other organizations, including the US government. BP, Transocean, and Halliburton organizations involved in a drilling operation, which went wrong. The accident makes an interesting case in decision-making involving different organizations, but also in the fact that the oil leak was so large it involved a number of states and the US government in the decision-making process. This is a prime case of multiple decisions made by companies, states, and the US government that occur together and do not produce a good effect. Also involved were other countries that offered help in the form of oil skimmers, etc. It is not proposed to examine the offers beyond the fact that they were made. The refusal of this help is quite interesting in its own light, especially in that it was between those countries and the US government!

It is appropriate before discussing the accident to discuss the drilling operation that led to the accident and the involvement of the different organizations in this process. BP was the client in that they were the owners of the site; the exploration rights were leased by BP from the US government. BP personnel were in charge of the operation, Transocean was the owner of the drilling rig and ship and provided personnel for these operations, and Halliburton was the provider of cement and drilling "mud"; these materials are used respectively to support the drilled hole and for the drilling process itself, as a lubricant.

#### 7.6.1.2   Accident Description

The BP-caused oil leak into the Gulf of Mexico was one of the single biggest releases of oil in recent time—some 4.9 billion barrels of oil were released along with methane gas. There have been many oil leaks over the years in various locations by different sources. Significant leaks have come from oil tankers that have been damaged at sea or by collisions with other ships or running into rocks. The Exxon *Valdez* was one such accident. The *Valdez* accident occurred in Alaska and released tons of oil into the bay, damaging animals, birds, and fish, while covering beaches with tar and oil deposits. The oil leak from the *Deepwater Horizon* rig affected a large number of states around the Gulf area from Texas to Florida. The impact of the oil releases in the Gulf is much the same as mentioned earlier.

The description here is intended to give an overall picture of the accident, and later the focus will shift to the measures taken by the parties to control the release of oil, terminate it, and minimize its effects on the neighboring states. The accident's real impact was on the long-term loss of fishing and vacation-related jobs. The Natural Resources Defense Council (NRDC) reported on the effects of the accident after one year (NRDC, 2011). The accident was both an ecological and economic disaster and is expected to have a long-term economic depressive effect on the whole region.

The oil rig was a ship with a drilling rig mounted on its deck. The rig crew had just celebrated "10 years of no accidents'" and managers from the various companies associated with this achievement were present for the celebrations before the accident took place.

The drilling operation was behind schedule and there was pressure to drill quickly through to the oil pool below. Some of the operations that went into preparing the hole depended on the use of special cement to constrain the oil paths and prevent the ingress of water. There was also the use of a number of spacers within the bore. It was said later that the cement was substandard and the number of spacers was fewer than should have been used. Clearly, this is a question of opinion. The spacing decision was taken to speed up the drilling process; again, this was an opinion. This will continue to be discussed.

The significant event was that during drilling there was a large release of gas, possibly due to solid methane being transported to the surface and then evaporating on the way up. From a safety point of view, one of the deck crew members failed to warn the ship's company of the problem. Then was an explosion that killed a number of crew members. This was followed by a large release of oil and the blowout preventer (BOP) safety valve system failed to cut off the oil flow. The BOP is a mechanism consisting of sets of valves, which are independent of each other and are closed by redundant ram drives, that can cut of oil flow, like closing gate valves. In addition to these cutoff valves, the BOP is equipped with a shear ram to cut the pipe leading to the ship and seal off the flow.

For this situation, the oil does not need to be pumped out of the ground. The pressure in the pool is very high, and oil just flows out under high pressure. Under these circumstances, the BOP valve system needs to close and shut off the oil flow to the surface and then all will be fine. However, all did not work properly and the result was 11 persons were killed and 4.9 billion barrels of oil were released into the Gulf.

### 7.6.1.3 Accident Analysis

It appears that the key event leading to the large uncontrollable release of oil was the complete failure of the protection systems to shut off the oil. The safety design appeared to have been designed with both redundant and diverse aspects. From a theoretical point of view, this was the correct approach. Accident analysis has told us that a mixed redundant/diverse combination is the way to go. The idea of using diverse elements is to obviate the impact of "common cause" failures. However, the designers seemed to have missed out here in that the effects of the explosion wiped out all controls and probably distorted the pipes. This common event then led to the massive release of oil into the Gulf.

Everyone asks questions about the sequence of the accident in detail, who was responsible, and subsequently who should be punished. Clearly, there was a number of factors at play—the design of the shutoff mechanisms, the quality of the concrete, number of spacers, pressure from local management, failure of the person to warn others, etc. BP did publish a report of the accident contributors from a mechanical point of view. However, it failed to consider the event from a total point of view, from design to final cleanup. But then that was not the direction given to the main research team. The previous brief description is sufficient to at least indicate the presence of certain actors. Clearly, BP is the ultimate holder of responsibility for causing the leak. There are other companies that played a significant part in the accident, but BP personnel were in charge.

There were things that BP appeared not to do, one of which was a risk-based study of what could go wrong, including the consequence of failure of the safety systems to perform as designed. Often, the designers of such systems are the last people to consider what can fail. One wants to move from success space to failure space and designers often cannot make that transition. It appears that the basic failure that led to the accident was the failure of the safety system to function correctly. Even if there were contributory influences that speeded up the event, like poor quality concrete, failure to deal with solid methane effects, and the BP supervisor pushing the drilling staff, we always come back to the failure of the redundant/diverse safety system failure as the main failure. In this, BP and the designer must take responsibility—BP for failure to review the design from a probabilistic risk point of view, including the risk to the company of its failure, and the designer for not reviewing system problems leading to failure to work, including loss of signals, actuator failures, etc.

However, once the release was under way, what should have been done and who should be held responsible? So now one moves into the release stage of the accident. People at BP, etc. will focus on the technical issues of the mechanisms associated with the stages before the explosions and the immediate actions taken by BP and others. In some ways more interesting from a management point of view is to diagnose the actions that BP, the Coast Guard, local authorities, and the US government took or did not take to respond to the release and the other issues that came up. The first thing to ask is—what are the responsibilities of the various parties?

From a legal point of view, it would seem that BP should be held responsible, since they started the process. However, the licensing authority has some involvement; by their actions BP was found to be a capable company to do the drilling, so by implication they should be held equally responsible unless BP was deceiving them in some way.

The US government is a key player here and has the responsibility to protect and defend the public. So surely they needed to have responded quickly, possibly through the Coast Guard, to assemble skimmers and other oil-collecting ships to remove the oil as quickly as possible, even using the US government's good offices to enlist help from other countries. They should have set aside trade constraints, which prevented offered help arriving—like setting aside the Jones act. So what is the difference between this situation and the response to various hurricanes and storms?

The US government continued to pursue BP to assemble a large group of ships to minimize the effects of the oil spill. It appears that the US government did not understand the limits of BP's capabilities relative to the requirements. Only the US government had the prestige and resources to respond to the accident in a timely sense and minimize the ecological and financial effects of the spill. Although BP is a large and wealthy company, its wealth is tied up in its oil wells, refineries, and personnel. Given time, they could have negotiated to sell resources, but the value of the resources would be worth less than their face value. Time was of the essence, and the US government waited too long to act and oil leaked ashore, doing damage to sensitive areas.

### 7.6.1.4    Organizational Analysis

Figure 7.10 depicts a VSM diagram with the BP company officials, the Obama administration, the licensing and leasing department of the US government, and the US Coast Guard. The companion companies involved in the oil exploration are subsumed within the BP organization. The oil licensing and leasing authority was called Mineral Management Service (MMS) and one result of the accident was to redefine the role of MMS. It is now called the Bureau of Ocean Energy Management, Review, and Enforcement (BOERF) and it covers more than leasing. The function of the Coast Guard is law enforcement on the seas and search and rescue. In the case of the post-accident situation, the Coast Guard was given the task of coordinating response to the spill.

The object behind using the VSM approach in this case is to diagnose the accident situation in terms of recovery actions. Figure 7.10 also shows the limitations of both lower and upper levels of the BP and associated organizations in not taking the correct actions to minimize the possibility of an oil leak or gusher, in this case. Within BP there did not appear to be the necessary resources to deal with such a major release.

The failure of the US government to act quickly to hasten oil recovery ships, pontoons, and portable oil barriers led to the large volume of oil released. Although BP is a large organization, its ability to react to the situation in a very large way is limited by its actual influence with other countries and organizations. The US government is in a much better position to do this. As far as resources are concerned, BP is a wealthy company, which does not mean that they can instantly have large amounts of cash on hand to pay for all of the rescue vehicles and personnel needed.

### 7.6.1.5    Conclusions

Figure 7.10 shows a number of different managers and organizations, associated with BP Oil Company as a whole, which covers exploration, refining oil, and distribution. This is the center of the company and the controlling body. Then there is the part of the company associated with exploration of oil properties throughout

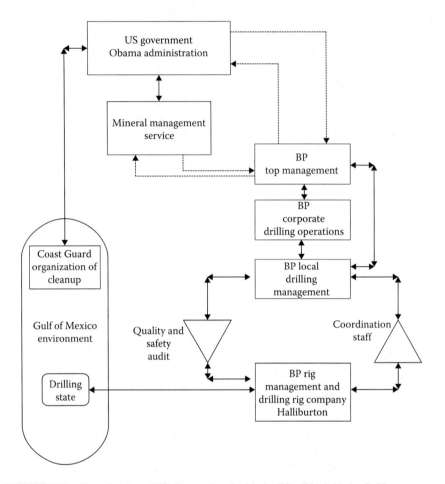

**FIGURE 7.10**   Organizations (VSM) associated with the BP oil leak in the Gulf.

the world and there is the organization that covers the drilling operation, and later it turns its job over to the exploitation group that collects the oil and passes it to the oil refineries.

The VSM model covers these operations to a reasonable degree. Included in the local drilling operation are the usual groups trying to ensure that the drilling operations are monitored, audited, and coordinated. Within these operations are persons responsible for the safety of operations. These are communications and actions within the BP orbit, as far as the drill operation is concerned. It is within these local operations that things failed. Advice from drilling personnel was supposedly overridden by the local BP project manager. Also, the person monitoring safety indications did not appear to observe warning lights indicating the presence of methane gas and alert the staff on the boat. Some 11 persons died and 17 were injured, because of the methane explosion and fire.

Further, it is believed that the fail-safe aspect of the BOP failed due to loss of redundancy, the failure of the backup isolation system to isolate the drilling tube and prevent release of oil. The precise mode of failures of the BOP has now been released (CSB, 2014). Clearly, the BOP failed and there were other contributory failures including some human- and equipment-related ones. In the design of things, the BOP failure is something that was not expected and if it had not failed, it is likely the topic would not be addressed here.

However, there were other contributions to the accident from other sources. For example, how was the frozen methane released and why did it cause an explosion? Why did the observer not give an alarm? The key issues appear to be the uncontrolled release of methane, the explosion caused by the release, and the apparent failure of the BOP device to act following the explosion. This left the oil to gush out of the well with no significant way to stop it in a short time. The actions of the BP manager to continue to drill may or may not have been significant; however, one is left with idea that some caution should lead one to hold off for a while until the situation is understood better. The ability of the frozen methane to pass through the drilling pipes, maybe due to poor cement, etc., is questioned. For example, do the missing spacers and poor cementation have any real effect upon the accident and its sequence and consequence?

Other organizations depicted on the VSM diagram also played a part in the accident. The MMS organization enters into the process by leasing the site of the Macondo well. In retrospect, the Obama government questioned what safety precautions were demanded by MMS and what regulations were required to be followed by BP (and other drilling operations) in drilling at these depths. One would have thought that it was in the interest of BP to have a clean operation, since the loss of 4.9 million barrels of oil at, say, $20 to $50/barrel represents $196 million to $490 million, depending on the price of oil at the wellhead. The loss of the well, the boat, and people is not insignificant and has to be added to the loss total. One also has to add to that the cost of cleanup.

A review of the accident indicated that the Obama administration or government agencies, such as the Coast Guard operating under administration's directives, did not act quickly enough to avoid the spread of the oil to the coasts of states from Texas to Florida. It does seem right to hold the US government responsible for it failed to act quickly to minimize the effects of the oil on the populations of the states, the flora and fauna, and the local business (fishing to tourism).

The conclusions from the accident are as follows:

1. BP underestimated the issues with the local management relative to taking actions not in the real interest of BP as a whole. Decisions were made to save relatively small amounts of money, while risking much more: lack of perspective on behalf local management.
2. BP main management failed to initiate a risk study to identify the consequences of a failure of the BOP valve isolation system. Superficial understanding of common cause issues led to a locally reliable system that failed to perform. Local BP management did not have an effective quality control program.
3. Local rig members seemed to be not as well trained in safety aspects as they should have been.

4. BP analysis of the possibility of stopping the leak along with others was too optimistic, leading to a failure by BP management to state the correct time needed to stop the massive leakage.
5. The US government was fixated on BP's responsibility in the case of the leak and failed to see what its proper role was. It acted much too late and even then did not fully commit both US and other resources. In fact, it seemed to act against the interests of the citizens living in the Gulf regions.

### 7.6.1.6   Organization (VSM) Comments

The VSM model (Figure 7.10) shown here seems to cover the main aspects of the various organizations involved. The main item not covered is the viability of the BOP system to stop the oil from gushing from the well. The connections between the various parts of the BP organization seem to be represented and some of the functions were carried out, but one must question the success of these communications, since the head of the BP seemed not to be too concerned with his responsibilities in this case. In the VSM diagram, a management block directly associated with safety of operations has not been drawn. It is noted that the person who led the scientific investigation of the accident has been appointed to manage safety of operations. One feels that BP should have investigated the consequences of BOP failure and what mechanisms could lead to BOP failure to achieve its mission. This review function could have been incorporated into an audit function, similar to the CNO of an NPP organization.

The VSM model includes the US government, the Coast Guard, and the MMS in its depiction. Clearly, the US government stepped in to assume a leading role, but in a regulatory mode, not as a proactive organization to ameliorate the effects of the accident on the population. It is interesting to compare the actions of the Japanese prime minister's role in the Fukushima accident with that of the US president in this case. Here the US president seemed to be more interested in punishing BP, not coming to the aid of the persons in the surrounding states.

The US government could have taken steps to minimize the oil releases from reaching the land and contaminating the shores of several states. However, the Obama administration attenuated the variety (Ashby) by not authorizing the Coast Guard to take necessary actions to check the oil distribution using all available resources, including specially designed foreign boats volunteered by their governments.

### 7.6.1.7   Postscript on the Macondo Well Accident

A review was made of the BP report on their analysis of the accident (BP, 2010). The report was very extensive and very technical, but somehow a little off course in that while it goes into great detail on the accident, it did not deal with the key elements of decision-making under high-pressure conditions and how to avoid human errors leading to a massive oil release. Interestingly, BP awarded the leader of the analysis team with the post of head of safety! It was a very detailed report and several organizations and experts were involved.

A review of the Chemical Safety Board's report (CSB, 2014) was made and it was very significant in that it revealed the series of failures that led to the BOP failure to shut off the oil. CSB was able to retrieve the BOP and examine it to see what could be determined about its failure to meet its design objectives. The basic concept of meeting a reliability goal by the mixture of redundant and diverse systems is good and this approach has been used in the nuclear power industry for about 45 years very successfully. The success of the safety systems in the nuclear industry is due to periodic testing coupled with care in maintenance. Many of the requirements placed on the design and operation of these safety systems were codified in Institution of Electrical and Electronic Engineers (IEEE) standards (see www.ieee.org).

The BOP had two redundant systems and a diverse tube cutoff system, which was the ultimate safeguard. All of these systems failed due to different failure modes. The redundant systems failed mainly due to a lack of maintenance. In one case, a check that a battery was operating correctly was not carried out and in the other case, a failure to check the connections to a coil. These were organization failures and could have been avoided. It is amazing that BP failed to ensure that procedures were written, trained on, and followed to ensure the system was periodically tested to ensure that the reliability of the systems were kept high. The failure of the guillotine cutoff system was unfortunate, possibly due to the malalignment of the drilling tube, possibly due to the explosion. Maybe the design needs to be reexamined, so that it can override these types of problems.

## 7.7  RAILWAYS

### 7.7.1  INTRODUCTION

Most accidents occur in the railway business due to railway configuration issues from cost considerations. For example, railways have accidents caused by trains running into each other, that is, trains being on the same track at the same time going in opposite directions. There are many examples of this both in the United States, France, and other countries.

The fundamental cause of these accidents is the fact that the cost of the lines and its preparation is high, so the railways try to run trains on the same rail to avoid costs. The single service line has bypasses every so often, so one train can pass another safely. Unfortunately, the signal and control arrangements, every so often, fail to alert the driver to use the bypass. Sometimes, it is the fault of the driver entering the section of rail on which another train has been given the right of way. He does this often by failing to see the signals informing him to stop! Some of the older lines used to have mechanical devices to automatically stop trains if the signal was up! One assumes that these systems went out of fashion due to overhead to get the trains going again, or was there another reason?

Other times the train's schedule does not match the schedule given to the line controllers, so the state of the line does not match what is needed. Examples of this are the Flaujac train crash in France, August 1985, between an express train and a local train (Carnino et al., 1990). Another is the crash that occurred in Chatsworth, Los Angeles, on September 2008 between a freight train and commuter train, Metrolink.

The other situation that leads to train crashes is when the train exceeds the speed limit for a bend. Again there are many examples of these: the accident (7 killed) in

New Jersey in May 13, 2015 (see Stolberg et al., 2015), and the accident in Northern Spain, Santiago de Compostola (Govan, 2013), where there were 77 killed and 143 injured.

One could argue that these accidents are the fault of the operator/train driver. However, one could suggest that the owners of the railways have some responsibility in the matter in that the rail systems are not designed to be accident free. What is meant by this? The single line approach for trains running in both directions will fundamentally lead to accidents, since there will always be situations in which either the automatic system or the human controllers will fail. The cost of designing the rail system like this will lead to human deaths and injuries.

The same is also true of bends. The rail lines could be constructed to avoid sharp bends, but trains have to negotiate built-up areas, where the cost of changing layout designs includes removing houses, and sculpting land is possibly seen as unnecessary.

### 7.7.1.1  Kings Cross Underground Fire, November 18, 1987

All railway accidents do not always stem from the same causes discussed earlier; for example, the fire that occurred on the London Underground at the Kings Cross station. This accident seems to be unique in that the fire occurred on the escalators going from the arrival level of the trains to the station exit.

The direct cause of the fire was due to passengers dropping their cigarette matches (still burning) on their way up the escalator. The lighted matches set fire to the excess grease used to lubricate the joints of the stairs. Unfortunately, the old grease was not removed during servicing of the stairs and there was a buildup of the grease along with fibrous materials. The use of matches at this time seemed to be related to the fact that smoking on the Underground was banned in 1985, so the passengers lit their cigarettes on their way out. One could expect that the matches would not cause a fire every time they were dropped.

A new phenomenon was noted as a result of the fire. The fire traveled up the escalator tracks and not straight up into the air, as expected. This made things so much worse. The treads of the escalators' steps were made of wood, so these caught fire. The smoke from wood and grease traveled along the escalators, filling the halls with smoke and causing particles to fall on arriving passengers. Some 31 persons died and 60 were burned or suffered from smoke inhalation.

The response of the fire services to the situation was slow, since they underestimated the conditions and thought that only a light smoke was generated, rather than the actual conditions of heavy smoke and fire.

### 7.7.1.2  Organization Analysis

The London Underground transport management did not clearly understand the probability or disastrous consequence of a large fire in the Underground. The failure to realize that the heavy grease and fibrous material was a source of combustible materials had been passed over for many years. Tests carried out later showed that the heavy grease was difficult to light, but much easier in combination with the fibrous material.

The tunnels at the Kings Cross/Euston stations had grown over time, but the issue of evacuation of passengers and staff safely in a timely manner seemed to have been missed. Equally, the fire services seemed to underestimate the consequences of fires. One wonders why tests were not done on the combustible materials in the Underground system to see what their impact on safety was. This seemed to be another case in which problems crept up on management, like the Aberfan coal tip disaster in Wales (1966).

### 7.7.1.3  Comments on Railway Accidents

This briefly covers the field of railway accidents and the relationship of decision-making to those accidents. The impact of capital expenditures on railway layouts was, and is, a deciding factor and determines the railway solutions. Also, it is not clear that any other solutions were considered. The whole rail design process seems to have developed without considering the implication of the effect of these decisions on safety of passengers or employees. This looks to be a carryover from an earlier time and the process has not been reconsidered.

In other aspects, problems like fire control creep up on management (London Underground) until management is forced to make decisions to fix something that was building up over time. So management decision-making is reactive and not proactive. Again, the correction of design and human reliability issues are reactive. For example, if there is a curve-safety issue, then the solution seems to be "Let's design an automatic system to prevent the operator going quickly around the curve" rather than redesigning the curve and/or the train suspension system to eliminate the problem.

## 7.8   NASA AND AIR TRANSPORT

### 7.8.1   NASA CHALLENGER ACCIDENT, JANUARY 28, 1986

#### 7.8.1.1  Background

Space exploration is a risky business for both man and robotic missions; witness the number of failed operations going back to the German rocket (V2) launches during World War II. NASA's record has been relatively good, although it could have been better. Prior to the *Challenger* accident, there was the Apollo 13 (April 11, 1970) accident involving the rupture of an oxygen bottle, and the oxygen-induced fire during takeoff preparations for Apollo 1 (January 17, 1967). NASA has been a recipient of great expectations with regard to the safety of astronauts. The overall failure rate is about in the range of 1 in 25 to 1 in 50 launches for the Shuttle (Orbiter), despite much care in preparation for the launches.

Safety, in general, is enhanced by the use of redundancy and diversity, much used principles in the nuclear power business. However, in the rocket field weight is of great importance, so one has to design highly reliable systems without recourse to multiple backup systems. This means that there is a limit to the reliability of Shuttle missions and any deviation from the basic systems design or operation can quickly reduce the safety of the missions. The *Challenger* and the later *Columbia* (February 14, 2004) accidents are related to the failure of management to fully understand the impact of design and operational limitations on the basic safety of the Shuttle!

### 7.8.1.2   Description of the Shuttle

The normal process of launch is for a Shuttle to be assembled in a large building (vehicle assembly building) and then transported to the launch site, where the preparations continue, including filling the main fuel tank and ancillary fuel services. The shuttle consists of three main components: the Orbiter, the main fuel tank, and the two auxiliary solid fuel boosters (SFBs). The connections between the fuel tank and the three main rocket engines are made early in the assembly process. Apart from all of the preparatory processes, which take a large amount of time, the launch process proceeds once there is a "go" from mission control. The main engines, liquid fueled, are started and when the thrust reaches a certain level, the SFBs are ignited.

The SFBs are constructed in a number of parts to assist in shipping. There were two versions of the bid for the SFBs; one was the then current design and the other was an integral unit. The bidder for the first type won. Perhaps the *Challenger* accident would not have occurred if the designer of the second design had won! The joint between the components was sealed by a set of two "O" rings and heat-resistant putty.

### 7.8.1.3   Analysis

Earlier launches had a history of bypass burns due to gases from the SFBs affecting the function of the "O" rings. The design of "O" ring installation is such that their flexibility is an important feature of their capability to seal the joint. Normally one would design the installation such that the "O" ring would be squeezed under normal conditions and once the back pressure occurs the ring would slide to completely seal the opening. Clearly, once the ring loses its flexibility, it then fails to completely seal a joint.

At the time of the *Challenger* launch, the weather conditions were cold to the point of being icy.

During the prelaunch state, evaluation of the preparedness for launch is made. This is a very formal process and during the *Challenger* launch engineering advice was given not to launch because of the cold weather and the fact that the launch had been delayed for some time under these cold conditions. In fact, there was a criterion based on the temperature of the joint. The engineering advice not to launch was based on experience with the "O" ring history and the fact that the joint was considered a Category 1 safety concern (Category 1 means that failure leads to the loss of a Shuttle). The decision-makers were under some pressure to launch, because of the "Teacher in Space" aspect and that Ronald Reagan was going to announce this to the nation. The decision-makers thought that it was important for NASA to be seen to be able to launch as and when required. So it approached the manufacturer for a separate assessment, and despite the prior engineering advice, the representative of the manufacturer said that it was acceptable to launch. It has been asserted that there was an implied pressure by NASA on the manufacturer to agree to a launch decision!

### 7.8.1.4   Description of Accident

The result was that the launch occurred and a short time into the ascent the "O" rings failed to prevent bypass gas flow. The hot gases from the SFBs impinged on the liquid fuel tank and this led to a large explosion. The *Challenger* mission exploded and the astronauts, including the teacher, were killed. The subsequent investigation stated that the accident was helped by a large wind-shear force and the Shuttle broke up in part because of the strong wind coupled with the tank exploding. The wind-shear force would have tended to distort the SFB structure, putting increased separation forces on the "O" rings.

### 7.8.1.5  Accident Analysis

Part of the problem was the design of the SFBs. During the firing of an SFB, the whole SFB assembly is affected by the firing of the rocket motor. It distorts and vibrates like a partitioned set of rods. Hence, the flexibility of the "O" rings is completely necessary in the face of the movements of the SFB elements. They have to effectively seal the joints and prevent the impingement of hot gases on the liquid fuel tank and some of the support members holding the tank, SFBs, and Orbiter together.

Clearly, this is a case of suboptimal decision-making. Of course, the president would like to announce that a teacher had been sent into space, but I am sure that he was horrified that NASA interpreted his priorities and views in this manner!

Note: Later Richard Feynman (Rogers Commission) stated: "What is the cause of management's fantastic faith in the machinery? It would appear that, for whatever purpose, be it for internal or external consumption, the management of NASA exaggerates the reliability of its product, to the point of fantasy." He argued that the estimates of reliability offered by NASA management were wildly unrealistic, differing as much as a thousandfold from the estimates of working engineers. The statistical number of launch accidents to date of the Shuttles is 2/33 (0.06).

The analysis tries to relate the various conditions and decisions to the potential for a catastrophic failure of the *Challenger* Shuttle. There appear to be two main features behind the accident: (1) the essential limitations/weaknesses of the "O" ring design of the joint and (2) the NASA management's basic belief that the failure probability of the Shuttle was very low (see the comment of Feynman). Given the implied pressure stemming from the president's interest in the "Teacher in Space" aspect and coupled with the belief in the robustness of the Shuttle, NASA disregarded the caution of the engineers not to launch as not proven and it decided to launch. However, NASA still needed to get some outside "expert" to confirm its opinion and found it with a Thiokol vice president. It has been stated that Thiokol engineers had also informed NASA of the incipient problem and suggested waiting until the Shuttle SFBs warmed up!

There are a number of decisions behind the *Challenger* accident, some associated with the conditions (icy cold weather) and others made much earlier during the design process and related to the cost evaluation stages. One could say that latter decisions dominate. Of course, there are dependences between the various decisions, for example, cold conditions and "O" ring flexibility; so if the joint design was not affected by weather changes, then the influence of the cold conditions would be small to nonexistent. We cannot say that other considerations associated the overall design would not lead to a loss of a Shuttle, as in the case of the *Columbia* shuttle accident. In fact, the later *Columbia* accident has very similar roots in the decision-making of NASA management in a number of ways. It was recognized that there were SFB design weaknesses and some personnel were aware of these in deciding to launch or not to launch.

Feynman in his analysis stated there were two items that led to incorrect decisions: an incorrect analysis approach that depended on the idea that a successful launch implies that all following launches will be successful, and that degradation of safety controls occurs over time. Both of these have been seen in other circumstances. All too often a situation deteriorates until an accident occurs, and then the management tries to reinstate the correct situation, sometimes too late!

The pathways through a decision process involving various parts of the equipment successes/failures can lead to an accident or success. As pointed out, the design features of the jointed SFB that could lead to failure of the launch are if the "O" rings are inflexible and one relies on the fact that the joints are flexible. High winds can also cause the joints to open and allow the hot gases to bypass the "O" ring seal; also cold weather can lead to "O" rings being less flexible, when there is movement of the joints. So cold and high winds can individually and together lead to problems with joint sealing. Depending on the wind forces, there might be problems even in warm conditions.

In the old days, when space travel was seen as even more dangerous because of engine explosions, the capsules had escape rockets attached to them in the case of main engine rocket failure! A redesign of the "O" ring joint seems to be the only clear way to avoid having escape pods and operate in reasonable high winds and cold conditions. The current design has increased the number of "O" rings to three; it is not clear that this would have been sufficient to enable launches to be independent of the weather! The Orbiter (Shuttle) program has since been closed down.

The following points should be made:

1. Design of the "O" joint, part of the successful design bid, was not sufficiently robust given its "Category #1" rating and NASA launch needs, that is, somewhat independent of weather.
2. There seemed to have been an incorrect prioritization of goals; safety should take precedence over other objectives. It has been asserted that there was a limited safety culture at NASA.
3. The management did not give credence to the engineering staff, implying a great deal of egotism of the part of NASA management.
4. Management did not seem to have an appreciation of prior issues with "O" ring blow-bys. The engineering staff was aware of the limitations of the "O" ring assembly for some time and nothing was done about a redesign!
5. The engineering staff may not have presented their conclusions on established facts, rather than opinions.
6. Management elected to launch despite the fact of the uncertainty associated with a Category 1 component/system, "O" ring, design.

### 7.8.1.6 Comments

The pattern of NASA management's involvement seems to follow the same steps seen in other HRO cases, that management thinks that it is more capable in even purely technical issues, than those reporting to them. Managers think that persons reporting to them are significantly less able to recognize key issues than they are.

In this case, the management seemed even to place pressure on their suppliers of the SFR to support their view that the "O" rings would survive the launch and retain the hot gases and prevent the impingement of the gases on the main fuel tank.

Both the managers and their engineering staff had access to the same facts that the "O" rings represented a risk, since in earlier launches the recovered "O" rings showed damage due to scorching and luckily had held together.

The management's view was that "O" rings would hold and the engineers' view was that the "O" rings would likely fail. This represented a risk to fly. The prudent option would be to delay and let the "O" rings warm, until they were within flexibility specification! Unfortunately, for the crew of the Shuttle, the engineers were right!

### 7.8.2 Tenerife, Canary Islands Runway Accident, March 1977

#### 7.8.2.1 Background

The accident occurred, which affected two Boeing 747s, at the Tenerife (Los Rodeos, now Tenerife North) airport, Canary Islands, Spain. The planes were diverted because of an explosion at Las Palmas caused by a terrorist group. A number of planes were directed to land at Tenerife and this led to some congestion on the taxiway of the Los Rodeos airport. The situation improved at Las Palmas and the planes were allowed to leave. The planes had to be "unpacked" before they could leave. Planes had to reverse taxi down the runway and then turn either on the main runway or turn off a slipway onto the taxiway. The KLM flight was the first and back-taxied down the main runway, followed by the Pan Am flight, except the Pan Am flight had to proceed down the main runway and then turn into a slipway/taxiway to the turnaround point at the end of the runway (see Figure 7.11). The problem was the Pan Am pilot missed the turnoff point and headed for the next slipway. The KLM plane was taking off, when it ran into the Pan Am flight taxiing on the same runway, but moving in the opposite direction. The majority of the passengers and crews in both planes died. Total fatalities were 583 killed and 61 injured. None on the KLM flight survived.

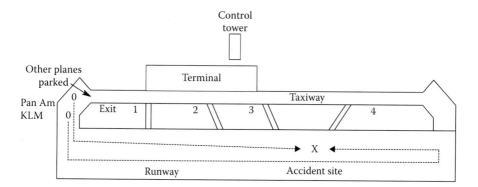

**FIGURE 7.11** Depiction of the runway in Tenerife including routes taken by the KLM and Pan Am Boeing 747 aircraft.

### 7.8.2.2    Accident Location

The accident location is important, since if the airport had been a large international airport, it is very likely that this accident would not have occurred. In addition to size, the staff at the smaller airport was not used to or trained to handle these numbers of large airplanes. Figure 7.11 shows the layout of the Los Rodeos airport. It is a small airport with a single runway and a taxiway. The airport was designed for regional traffic to and from Spain. It was not designed to handle 747s, especially a large number of them! All of the 747s were parked in the left side of the runway.

### 7.8.2.3    Accident: Sequence of Events

Both planes were operating under the control of the flight tower personnel. Because of the stack-up of flights redirected to Tenerife, the normal access taxiway was unavailable. The flight tower directed the KLM plane to proceed via the runway to the far end and turn around. Prior to the plane proceeding up the runway as directed, the KLM chief pilot had the plane filled up with gas to save time later on. Once the plane reached the end of the runway, it turned around and waited for orders on when to take off. At much the same time, the Pan Am 747 captain was also ordered to proceed down the runway, but was told to take the #3 slipway and then proceed to the end of the runway. At first, all was going well. The KLM flight was in position and Pan Am was moving toward the slipway.

There were communication difficulties between the tower and both sets of planes. The Pan Am crew was told to take the third slipway onto the taxiway. The Pan Am pilot appeared to have missed the third slipway and proceeded toward the fourth slipway. About the same time, the KLM flight was receiving clearance on what to do once it received instructions to take off. However, it did not receive the go-ahead. There was some confusion at this point and the KLM flight thought that it had clearance to take off. The KLM crew did a talkback to confirm the takeoff instruction. In these communications, there seemed to be problems between the KLM crew and the tower. The parties did not use standard statements, at least according to later investigation reports, and this eventually led to the accident.

In addition, it was reported that there was a "heterodyne" beat caused by simultaneous signals coming from the Pan Am crew. This also added to lack of clarity in the instructions among all parties. This beat signal led to the KLM flight crew not hearing the control tower statement and also failing to hear Pan Am crew's statement that they were still traveling down the runway! On top of all of this the weather conditions were poor and none of the crews or tower personnel could see each other! Despite the fact that the flight engineer on the KLM queried the captain that the Pan Am may still be on the runway, the captain took off. Just before the collision both captains saw the other planes and took evasive action, but it was too late. Pan Am moved off the runway and the KLM tried to lift off.

Figure 7.12 shows the events that together led to the fatal accident. Figure 7.13 indicates the pathway of how the accident occurred. There were several opportunities during which if alternative actions were taken the accident would not have occurred. Two of the events are associated with decisions made by various persons and the other event deals with communications difficulties. Diverting the

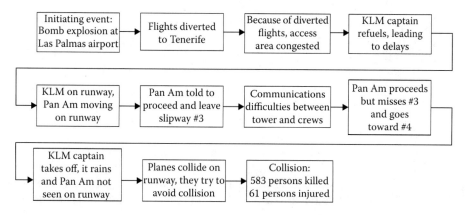

**FIGURE 7.12** Event sequence diagram for the Tenerife accident.

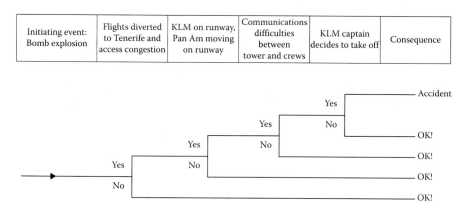

**FIGURE 7.13** Sequences of events for the Tenerife with alternative paths.

flights may not have been possible, but it appears that the Spanish air traffic control involved did not appear to consider all of the consequences of landing 747s at a regional airport!

### 7.8.2.4 Analysis

The following points should be made about the accident:

1. Following the rescheduling of the planes, the local management should have reviewed the state of the airport, how the arrival of the 747s could be accommodated, and what problems could occur. This last point is the key one.
2. Stacking of the 747s impacted the availability of the taxiway.
3. The airport management should have carried out a quick risk–benefit analysis to understand what the effect was of allowing only one plane at a time on the runway system compared with allowing two planes to move at the same time.

4. The management should have realized the consequence of language difficulties (American, Dutch, etc. personnel versus usual Spanish speakers) on
operations.

5. If a two-plane moving operation on the runway/taxiway was going to be
used, an additional person or persons should have been positioned on the
taxiway to ensure no planes were on the runway at the time that a plane was
taking off. This meant that there ought to have been a management control
set up to coordinate and control movements.

6. There was no redundant or diverse controls set up in the case of the operations. The management seemed to think that the prior method of operating
the airport did not need to change in the face of having a number 747s flying
in. Remember that the airport was a commuter airport (infrequent operations and small planes).

7. The controlling authority for control of the airspace should have taken
steps to assess the situation before allowing 747s to be directed there. This
includes suggesting changes in operational procedures to the airport management. It is not clear who the controlling authority was. The probability
is that it was the Spanish equivalent of the FAA.

8. One can see from the previous comments there was a severe underesti-
mation of organizational difficulties and therefore the prime blame for the
accident should have been placed on the management of the facility (local
and overall). Planning and organizing of the plane movements should have
been up to the management. Communications difficulties were just the
instruments by which the accident was promulgated.

9. The KLM captain should have been more aware of actions of others, that
is, movement of Pan Am on the runway, and allowed for additional checks.
Maybe he was conscious of the time slipping away and this was having an
effect on his decision-making.

10. Radio communications between 747s appear to have been bad. This raises
more concern about the lack of positional information being generated.

### 7.8.2.5  Comments

This is yet another case where overall management control was not effective. In this
case, the bomb threat caused a degree of chaos and failure to ensure that the 747s
could be safely controlled when maneuvering on the ground. Clearly, there was no
thought given that things might not go smoothly and that extra care was needed: nonstandard working conditions, not being used to maneuvering big airplanes, language
and communications difficulties. They should have called a stop, if communications
were bad, and moved one plane at a time and given up trying to save time and fuel
by moving two planes at the same time.

## 7.9  ANCILLARY SAFETY-RELATED INCIDENTS

A set of incidents, near-accidents, and problems associated with design issues
are discussed in this section to further illustrate problems related to management
decision-making. Not all incidences lead to major accidents. The incidents discussed

here are associated with the nuclear power industry. This is because the authors are closely associated with it. One can be sure that similar incidents can happen in other industries, because the underlying processes for the training of management and developing their experience base is very similar. We believe the training approach in the military is a little different for obvious reasons; the risks for the participants are much higher and they need to be very well prepared to make good decisions, especially if they are to succeed operationally and live.

Sometimes, even regulatory bodies can perceive safety issues developing with an organization without being able to take an action to resolve the problem before it leads to a severe accident. Another case is what seems to be a straightforward mechanical cleaning operation that leads to the release of radioactivity that harms maintenance personnel, due to design failures unforeseen by management. The examples given here are from different countries and help to illustrate that problems are not isolated situations limited to just one country.

The descriptions given here will be brief with the emphasis on the interpretation of the event as far as decision-making is concerned. References to more detailed descriptions of the incidents are given, in case the reader might need to examine any example more closely.

The general conclusion is that an involved and educated management is better prepared to deal with incidents/accidents and reduce the impact of accident progression from start to finish, thus reducing accident probability and its consequence.

## 7.9.1 NPP Containment Sump Blockage

The list starts with an example where a research team was approached by an international regulator about a potential NPP safety issue (Lydell et al., 1986). The issue was, can the operators recover from a sump blocked due to the insulation being blown off the walls of the containment? The team was asked to evaluate the capability of control-room operators to identify the problem and operate the equipment— pumps/valves to reverse flow to blow out the insulation material that plugged the sump. Failure to recycle the sump water to cool the core could lead to the destruction of the reactor core.

This was initially regarded as a human reliability problem and the team delved into examination of the location of valves and controls and the capability of the operators to understand the need and to work out what to do and how to do it. The team reevaluated the problem and decided that the significant safety issue was, was it even possible to clear the debris from sump by reversing the flow? It was recommended that the regulator carry out a series of tests to see how the insulation would break up, drop into the sump, and then be pushed out by the reverse flow. It was realized that for some conditions it could be very difficult to unblock the sump and this could lead to core cooling problems. The issue therefore was not a case of whether the operators could carry out a maneuver, but whether it was even possible to restore cooling. Clearly, this was a significant safety issue. Following our report, the regulator was too busy to deal with it at the time. He was caught up with international concerns about severe accidents and how to recover from them!

Later, it turned out that the plant that the team had considered in the research study was involved in an incident very similar to the one considered. Luckily, the plant was operating at low power when the containment insulation was blown off and sump was clogged. The operators were able to recover from this event. If this had been not been the case, there could have been a major accident and as covered in the previous "major accidents," the result could have been similar to TMI #2 unit.

Later, since a core melt situation event occurred, it caused the international communities to react and become involved in studying the plugging of reactor sumps due to containment insulation becoming detached. Many large research investigations were carried out.

Subsequently, the decision was taken to close down the plant, since it was considered too unsafe. People in the nearby city and country (Denmark) had been saying that for a long time and had been pushing to have the plant shut down. It really was unsafe, for a short time. If an accident had occurred, it would have been an economic disaster: cost of destroyed plant, replacement power, and cost of long-term cleanup.

So what is the lesson? Studies are theoretical, but truth is often confirmed by an accident actually occurring. Unfortunately, the position taken by some managers is, if it hasn't occurred, it is unlikely to occur and one is worrying about things unnecessarily. This appears the first law of management! Don't concern me now, tell me when it is about to occur. The second law of management: blame your people for not telling you; it was, and is, their fault! Of course, we disagree with both of these statements. A prepared and concerned management would have taken action, once the situation was revealed to them. It is better if the evaluation had come from the manager, but failing this, it is just as good if it comes from an advisor and the manager accepts the evaluation!

The wonder is why this incipient state had not been recognized before. Surely, insulation had become detached earlier? It is that either no one had considered it, or if someone had, it was ignored! Clearly, once it moved from the theoretical to the actual, the world paid attention.

### 7.9.2 Hungarian VVER Fuel Cleaning Accident

In preparing for the possibility that an accident might affect your plant and what actions you should be prepared for, you tend to prepare for the more usual accidents that have occurred in the past. Remember the idea of the design-basis accidents. Experience has shown us that accidents can be complex and involve combinations of natural events along with both equipment and human failures. However, accidents can arise from other sources. One such incident is covered below.

A serious accident occurred at Paks unit 2 VVER (Hungary) in April 2003, with the release of radioactivity. The accident involved the case of fuel being cleaned in vessels specially designed for this purpose. The partially used fuel was taken out of the core for the purpose of cleaning the surface of the fuel that was contaminated by magnetite (iron oxide) from the SGs (HAEA report, 2003). The fuel and some reactor internals were placed in the vessel for this cleaning purpose. The cleaning system was designed and built by Areva/Siemens.

The cleaning operation appeared to be going well, with the fuel elements reaching their cleaning time and the maintenance crew waiting for the reactor internals to be completed, when a radioactivity release was detected. The report covers the subsequent actions by the maintenance crew, which included partially opening the clearing vessel to examine the state of the fuel and led to a further release of radioactivity in which some of the crew were affected. The reactor building was evacuated. Subsequentially, the head of the cleaning vessel was removed and it was found that most of the fuel was damaged and some pellets (fuel) were released. To prevent the possibility of the pellets leading to nuclear reaction, borated water was added to ensure that this did not occur.

The incident was eventually judged to be an International Nuclear Events Scale level 3 event! Radioactivity was released and it is believed that a maintenance man was affected.

The real question is, how did such an event take place? An earlier test of cleaning a small number of fuel elements (seven) was successful, but it appears from the investigation carried out by the Hungarian Atomic Energy Authority (HAEA) that the basic thermal design of this main system was faulty and that the designers focused on the cleaning process, rather than the whole design of the system. The report specifies issues associated with the internal layout of the fuel relative to the flow through the vessel. Also, some of the design characteristics of the fuel itself were not factored into the fluid/thermal design. It appears that neither the designers of the cleaning process nor the Paks management were sufficiently concerned about the safety implications related to the cleaning process. The implications related to decay heat removal have been a problem in the industry for many years. Here the Paks organization should have carried out an independent safety study of the process, including running tests to check the flow and temperatures. As an organization, they have been very concerned in the past about the safety of their reactors and have carried out many probabilistic safety studies under different plant operating conditions. The tools and methods that they applied in those cases ought to have been used here, especially the idea of being concerned about safety and dealing with fuel having a degree of decay heat from the prior reactor operation. They may have fallen into the trap of relying on the expertise of a third party. Ultimately, the safety of plant operations falls on the shoulders of the management of the organization!

### 7.9.3 San Onofre NPP: Replacement Steam Generators

This is an interesting case of how design problems with replacement SGs stemming from management inadequacy led to SG tube failures, radioactive releases, NPP shutdowns, and compliance problems with the USNRC (see Joksimovich and Spurgin, 2014). To cap this, conspiracy charges may be brought against some members of Southern California Edison (SCE) and the California Public Utilities Commission (CPUC).

The sequence of events that led to the release of radioactivity is as follows: SCE decided to replace the current SGs and replace them with SGs that hopefully would have a longer life than the ones being replaced. Mitsubishi was contracted to build the new SGs and they were installed and were operated for a short while.

The earlier SGs for NPP units 2 and 3 were replaced at different times. A tube leak occurred on one new unit #3 SGs; radioactivity was released (January 31, 2012) after about 1 year of operation and the plant was safely shut down. Usually, if a tube leaks, it is plugged and the plant continues to operate after investigation to see if other tubes are involved. The other tubes are plugged, if the wall thickness does not meet NRC tube wall thickness criteria. Multiple tube investigations were carried out and it was found that some 807 tubes did not meet the NRC criteria and were plugged. Unit #2 was then inspected and it was in a similar condition: 510 tubes were plugged. The USNRC said that unit #3 was the worst record in history and the unit #2 was second worst in terms of number of tubes plugged at a given time. The next worst was 107 in 5 years (South Texas Project). The others in the record are much less worse than this (see table in www. NRC.gov/info-finder /reactor/songs tube-degradation.httml# data, plotting number of tubes plugged versus plants and period of operation before plugging).

This pointed to the inadequacy of the new San Onofre SG design and the rest of the problems started, which involved the NRC and decisions about what to do about the relatively recently installed SGs. The options were plug the tubes and run at part load, shut down the plants and replace the SGs, or completely shut down the plants.

However, the interest here is in the impact of design decisions on SG performance. Steam generators have a finite life, depending on a number of factors. These factors are tube corrosion and wear coming from tube vibration due to hydrodynamic forces. Work has been going on in the industry for some time to increase the life of the SGs to be more compatible with the life of the nuclear plant itself.

What seems to have happened is that SCE management did not appreciate that changing the design of the SGs, in the hope of increasing the life of the SGs, was not easy to do. Their engineers along with others from Mitsubishi made large changes to the SG configuration from the earlier SG design. Their focus was on trying to increase the SG life by changing the tube materials to reduce corrosion effects, increasing the number of tubes, and also making changes to the SG internals, but in so doing they made the situation much worse, as evidenced by the tube leakages and plugging. Among other things, they failed to get advice from the industry research organization, Electric Power Research Institution (EPRI). EPRI had been funding research into SGs for many years and were particularly aware of the effects of hydrodynamic forces on SGs that can lead to tubes rubbing against each other and their supports and producing leaks! They were studying what has to be done to extend the life of SGs.

SCE went ahead and replaced both NPPs, with Mitsubishi constructing the SGs to SCE design. Mitsubishi did not have any experience with the size of San Onofre SGs. Mitsubishi built many SGs that were originally designed by Westinghouse and were smaller. The NRC stated after their review of the situation that the computer codes used in the design process were inadequate for the purpose. The original SGs installed at San Onofre were designed by Combustion Engineering and lasted for about 16 years. This does not seem to be as good as the unit #1 SGs, designed by Westinghouse in the 1960s, which lasted for 24 years, so it may be more difficult to extend the life of these larger SGs. Westinghouse plants of about the same power level have four SGs versus San Onofre's two SGs per NPP, implying that the tube surface area was twice that of Westinghouse SGs.

SGs are complex in their internal flow paths and interactions with the tube support structures. In fact, the elastic-flow interactions in SGs are more complicated than that of structural-flow interactions of air flowing over aircraft wings, which are thought to be complex! The SG supports are required for stability of tubes during operation, to avoid tube-to-tube impact, but must allow for thermal expansion of the tubes from cold to full power conditions and avoid inducing stresses in tubes. Supports must avoid impingement effects on tubes, and inducing cracking of tubes. The supports are there to help control the movements of the tubes so that the forces induced by the hydro-elastic effects do not cause tube wear.

It appears that the SCE management did not act prudently in their decisions related to the design of the SG replacement. In areas where one's explicit knowledge and training is limited, it is better to take a conservative approach and not make large changes in the design of equipment, since it may be more complex than it appears. It is also highly recommended to take the advice of experts in the field.

### 7.9.4 NORTHEAST UTILITIES: IMPACT OF MANAGEMENT CHANGE

This section deals with operational problems that occurred in plants but did not result in any significant accidents, but did affect plants' operations and their availability. One is always interested in experiments, since one can observe the effects of change. This section depicts the change that came about in the operation of a set of nuclear plants in the northeastern United States. The information contained here has been derived from two references: (1) MacAvoy and Rosenthal, 2005, and (2) Perrin, 2005. The research undertaken by MacAvoy and Rosenthal and Perrin was quite different, but together reinforces issues associated with the effects of the management change on the operation of the plants. Also one of the current authors was a human reliability assessment (HRA) consultant to Northeast Utilities (NU) in the period prior to the change in the upper management and can appreciate the effect of the change on plant operations.

The point of examining these operations is that one can see the direct impact of changes in management on the effectiveness in plant operations. MacAvoy and Rosenthal focused on the overall plant operations and the availability of plants relative to actions taken by management relative to staff. The period of their study was from 1986 to 1996. They also followed the actions of the USNRC responding to equipment availability requirements vis-a-vis requirements spelled out in the plants' final safety reports (FSAR).

A word here about the new management's philosophy, since it has a strong impact on the subsequent actions taken by management. The president was B. Fox and the CEO was W. Ellis. A report prepared by McKinsey suggested that there would be moves to lower the cost of electricity due to the entry of low-cost suppliers, such as units using gas as the fossil fuel. Ellis was previously employed by McKinsey and had acted as a consultant to NU. The philosophy of Fox and Ellis seems to have been heavily influenced by the McKinsey report. The state public utility commission did not think that it was likely to occur soon and that the utility was overreacting to the possibility. It appears that NU's management tends to follow the principles outlined in Chapter 13, in which shareholder value is a driving concern for management.

The president and CEO decided that overall costs of operating the plants had to be reduced and presented their plan, "A Strategy to Meet Competitive Threat," in 1987. This called for a reduction target of 13% below projected costs for 1990, with a 7% reduction in operation and maintenance costs and a 13% reduction in nuclear engineering and operations costs, (MacAvoy and Rosenthal, 2005). How this was going to work is seen by Mr. Fox announcing that some 1,500 workforce positions were to be reduced in 5 years!

The other cases that we have considered have devolved around accidents; in this case there is no critical accident. The effect of the management was to induce changes that were gradually deteriorating the plants' performance over a period of time. However, Perrin pointed out that a specific maintenance operation could have resulted in a serious accident induced by the operators working on a critical valve, "event 442" referred to by Perrin. The deterioration in performance stemmed from the actions taken by management to reduce staff and being too heavily focused on costs. Plant availability fell from about 90% to 56%. Plant shutdowns occurred due to equipment problems induced by failure to service equipment. Also, there were deficiencies in following up on FSAR requirements related to noncompliance of the plant components and systems. Again, these problems were associated with shortage of staff.

It could be said that the plants were very much approaching the point that a severe accident could have occurred due to issues related to personnel, systems issues, and operational problems. The staff reductions due to early retirement and layoffs ended with the skill-base of the plant staff being gutted; reductions led to the loss of supervisors and managers. The staff also informed the top managers that the plant safety was being impacted by their decisions, related to delayed maintenance and resulting equipment problems.

This information was ignored and conflicts grew between management and staff. The NRC became aware of these issues by whistle-blowers and was concerned about its impact on plant safety and whether there was sufficient protection for the whistle-blowers from management action. The usual management approach to problems is to "kill the messenger."

By 1994, considerable deterioration had occurred in the five NPPs within the NU portfolio. The average capacity factor had fallen to 57.6%. This was at a time when the industry availability was more like 90%. The deterioration in plant performance had reached the point that, in 1995, both the NRC and INPO, an industry body, had meetings with the board of trustees, rather than management, to draw attention to the availability issues with the plants. Nothing changed as a result of these meetings. Even the NRC's executive director met with the board; nothing happened and the NU management remained in place.

In 1996, the NRC designated NU a Level 3 company on the Systematic Assessment of Licensee Performance (SALP) and ordered the three Millstone plants shut down and not to started up until reconstructed and relicensed. Once the NRC became deeply involved, more issues became exposed. Compliance issues with the plant

FSARs became visible when the Unit #1 plant full core removal was found to be outside of the FSAR (this was brought to the attention of the NRC by a whistle-blower)! All hell broke out, as the NRC investigations went deeper, with a review of lack of an effective action correction process, as to condition of the NU plants. The story goes on—the line management did not respond to the quality assurance organization (QA) and root causes.

Later, it was stated that the objectives of the Performance Enhancement Program that had been agreed to by NU management, in response to earlier NRC requirements, had not been met. Review of the plants' state found many instances of problems stemming from failures due to lack of staff in engineering, which led to decreased reviews, and a lack of understanding of FSAR requirements. Some of these are the result of elimination of some engineering supervisory and management positions. Shortage of capital and staff meant that things did not get fixed. The staff tried to keep the plant working by "workarounds." In the case of safety systems, this approach is not acceptable! The lack of knowledge of the FSAR requirements was at the heart of the problems described by Perrin.

Eventually, the trustees got the message, but it was too late! NU did not survive as a nuclear utility and the NPPs were sold. However, the top managers were never punished for their activities. The policies of NU management led to the destruction of the utility. Many of the participants cannot be congratulated for their roles in this case:

1. The board of trustees failed to realize what the NRC and INPO were saying and did not take steps to change the policies of Fox and Ellis and fire them for their activities.
2. The NRC failed to act sooner and believed the words of Fox relative to carrying out changes defined in the performance enhancement plan (PEP) program.
3. INPO's strength was seen to be limited in terms of redirecting the utility management's actions.

The idea of a utility pursuing high-efficiency operations is not the issue. NU management undermined that idea by the way they did it. Just cutting personnel can never be effective; the operations need to be planned to achieve the reduction of organizational and maintenance (O and M) costs, without reducing plant safety and economics by maintaining the plants running at full efficiency. This is not an easy thing to do because of the safety requirements of NPPs. The utility was lucky in that the plants were not directly hit by a severe storm, like *Sandy* (2012), and experiencing an accident similar to Fukushima during this period, although the maximum floodwater height was about 10 feet rather than 48 feet. The important part is, how prepared are the management and the plant? Questions have to be asked about security of the power supplies: standby power (diesels), batteries and instrument power relative to the effects of flooding and the power of the waves to burst open watertight doors, etc. An interesting fact turned up during *Sandy*: sick and ill patients had to be moved from some six local hospitals to other hospitals because the emergency services (standby diesels) were flooded and not available after the loss of main power supplies!

These are examples of management failing to have the required knowledge, training, and experience to deal with these issues. They seemed to be completely unprepared to be able to achieve their objectives. There is a difference between running an organization and being an agent of change. Nuclear plants, by the nature of their enhanced safety requirements, have higher O and M costs than fossil plants!

# 8 Lessons Learned from a Series of Accidents

## 8.1 INTRODUCTION

A series of accidents and incidents were discussed in detail in Chapter 7 and the lessons learned from these accidents will be discussed here, principally from the view point of organizations and the roles of management in the safety and economics of operations.

Chapter 7 was devoted to analyzing how accidents arose and what the managers' part was in both causing and dealing with the consequence of the accidents. The accidents covered a range of industries and different situations. The industries were from nuclear, oil and gas, space and aircraft, chemical, and railways. Additionally, a later section of the chapter covered some interesting accidents or near accidents, which could be considered precursors of more serious accidents, such as containment sump blockage, fuel cleaning, replacement of steam generators leading to the shutdown of the plant, and lastly the impact of management changes on the efficiency of plant operations and the shutdown of the utility and nuclear power plants (NPPs).

All of the accidents and incidents yielded consistent messages relative to the operation of high-risk operations. The message drawn from these histories is that the quality of management needs to be improved. All of the industries involved need to take seriously the lessons learned. The accidents did not result from the industries being unlucky or that the accidents were inevitable. There are steps that can be taken to improve the situations and not just brush off the ideas by saying that accidents are inevitable and we cannot do anything about the probabilities. Yes, accidents may occur, but on close examination one sees opportunities to either stop them or at least mitigate their effects.

The usual action is to consider accidents as unlikely, or just blame the operators for not taking appropriate actions. Examination of certain accidents reveals a different interpretation of causality. For example, the Three Mile Island Unit #2 accident reports did point toward operator error. However, on closer review even the industry said the problem was with the training of operators and came up with a set of improvements: simulator training, better procedures, and improvements in instrumentation. So, one cannot say that cause of the accident was human error. Clearly, the actions were taken by the operators, but was this truly human error? In our view, the problem was with the industry as a whole. There was a failure to appreciate the complexity of the system and to understand the role of operators in controlling the plant under all conditions.

This scenario seems to repeat in most, if not all accident situations. In the analysis of organizations for each accident, the deficiencies have been pointed out. The deficiencies mostly rest with management, since they are empowered to make changes and make decisions to improve situations. These things do not rest with operators and others. This is not their role.

## 8.2   LISTING OF THE LESSONS LEARNED FOR EACH ACCIDENT

A listing of lessons learned is given in the following sections for each of the accidents studied. Some of the lessons draw from one accident may equally apply to other accidents. Organizational shortcomings are not confined to certain individual organizations, but are the manifestation of technical and managerial cultural limitations, which seem to be universal. For example, management seems to think that they need not be aware of the details of an operation, yet they are competent to decide the level of funding to carry out the operation. The same goes for determining manning schedules and selection of the skill level of persons needed to perform a task or tasks.

### 8.2.1   THREE MILE ISLAND UNIT #2 ACCIDENT

The Three Mile Island Unit #2 (TMI #2) accident was a wake-up call for the nuclear industry. Although the industry paid attention to many aspects of safety and considered the impact of accidents on the population, it is thought that methods and approaches together were not sufficient to ensure safety.

Some things became clearer following the TMI #2 accident:

- The knowledge and training of managers to be able to operate nuclear plants was inadequate, as opposed to operating gas and coal-fired plants.
- The importance of decay-heating was underrated by the industry.
- Total reliance of the industry on the control and protection systems to ensure NPP safety was a faulty decision. Further, it was coupled with a failure to realize the key need for trained operators to control some aspects of accidents not covered by automated systems.
- Failure of industry to realize that design-basis accidents, used for design purposes, do not cover all accidents.
- Failure to understand that for operators to act efficiently, they needed better procedures and that they should be trained in their use.
- Need of full-scale plant simulators to enable better training of operators was missed. Also, simulators were needed to show how better procedures could be integrated into the approaches to prevent or mitigate accidents

### 8.2.2   CHERNOBYL ACCIDENT

It is almost impossible to identify all the lessons learned from this accident; it was a fiasco of an accident. There were problems associated with running a difficult experiment without control over all the features of the experiment. It is extremely difficult to run off-normal experiments at an operating plant.

- Do not run difficult and sensitive experiments at operating plants without absolute control over the arrangements of the test.
- Do not change a trained test operator for an untrained operator just before a critical test is to be undertaken.
- Be aware of the dynamics of the reactor, under all conditions, especially for a reactor that can be unstable.
- Be sure that the benefit from running a test is of positive benefit and of little risk. The risk consideration should include an assessment of the probability of core melt or destruction of the plant. Here the benefit of carrying out this test was in our estimation never worth the risk.
- Buy reliable diesels from other countries, rather risk core damage.

### 8.2.3 Fukushima Accident

- To management, be courageous: even if the probability of a tsunami is 1 in 1,000, look at the possibility of the loss of the company, if it does occur. In the case of the TEPCO organization, the advent of a beyond-design basis tsunami did occur!
- Even if one hopes that an event will not occur, prepare yourself and the organization for the worst conditions. Follow the Boy Scouts motto: "Be prepared" and develop procedures to cover the worst combination of events and train the responders in their responses.
- Develop organizational structures that can respond quickly to events ahead of time, not during an event.
- Omit long lines of communications between headquarters (HQ) and field in reporting events and deciding actions. Concentrate on the organization being able to make decisions locally by providing guidance ahead of need and empower the local personnel to take actions when needed.
- Review all safety-related actions, like hydrogen gas release, and determine when the best time is to take the action to prevent increased damage to the plant or hazard to the public.

### 8.2.4 Bhopal Accident

- Organizations should not leave highly dangerous chemicals, capable of being released, without keeping close control over them at all times. Responsibility cannot be given away. All parties should agree on the required controls.
- If a company is no longer interested in the operation of the plant, it should be placed in a condition of not being a safety hazard to the public.
- Until a plant is safely shut down, all safety systems should be functional and tested to ensure operability.
- If the plant and its products represent a serious hazard to the public, then there should be an exclusion zone, which should be maintained, and public encroachment should be actively prevented.

- All parties to an operation like the Bhopal facility need to be closely involved with its operation, since all are responsible for its safe running.

### 8.2.5  BP Oil Refinery Accident

- All parties to an operation like the Texas refinery need to be closely involved with its operation, since all are responsible for its safe running.
- Maintenance operations need to be properly funded. Refineries acquired from other organizations need to be specially reviewed by HQ to ensure that their funding decisions reflect the real state of the plant and are within the capabilities of the current personnel.
- It is suggested that the HQ safety personnel perform a risk assessment based on both the physical state of the plant and the knowledge and training of site personnel. This is absolutely needed in the case of a plant taken over from another organization.
- Test local management to examine their plant knowledge and location of areas of the plant operation that represent the greatest risk for destruction of equipment, and death/injury to personnel/public. Pay close attention to the possible exposure of non-operational staff to the effects of accidents.

### 8.2.6  *Deepwater Horizon*/Macondo Oil Release Accident

- Organizations need to be aware of complacency. Successful operations can slide into accidents because of complacency and lack of control over details.
- Conceptually, the blowout preventers (BOPs) met the requirements for safety devices: redundancy and diversity.
- Safety equipment needs to be carefully monitored and tested. Even redundant systems can lose their redundancy if the batteries are not checked for charge or wiring isn't inspected to ensure coils work, etc. (These were the two types of failures that were discovered postaccident and affected the redundancy of the BOP system and caused it to fail.)
- Diversity can be impacted by failure of the device to function when needed. Bent or extra-thick pipes can lead to the device not working. The energy of cutoff drives should able to perform operations under most pipe conditions and be tested off-line to determine efficiency and capability of performing these operations.

### 8.2.7  Railways Accidents, Including Subsurface Railways

- If the railway management wishes to improve the safety record of railways, they should then redesign the foundation of rail systems; this includes reconsidering having trains operating on single lines and increasing the radii of bends, with the possibly of banking them.
- It appears that sophisticated signaling systems and driver education do not reduce the number of accidents. The old system of mechanical interlocks

seemed more successful. Perhaps the applicability of the safety schemes needs to be rethought.

- Underground rail systems need to pay attention to use of flammable materials in construction to avoid fire problems. The halls need to be clean and free of such materials.
- Consideration needs to be given to the timely evacuation of passengers following fires, floods, and other disasters.
- Fire brigades need to be prepared to deal with all kinds of smoke and gaseous releases.

## 8.2.8 NASA Challenger Accident

- The reliability of rocket launches is increasing, but this is due to organizations paying close attention to the design limits and keeping to the conservative side of operating conditions.
- The safety of rockets rests on the reliability of Category #1 equipment; there is no redundancy or diversity due to weight considerations. The management should not ignore the safety requirements and operational limits of Category #1 equipment in order to ensure the equipment will function properly during launch and post launch.
- There was an unwritten rule about Category #1 components—do not move out of their design range. NASA management moved out of the design range by overestimating the sealing flexibility of the solid fuel booster (SFB) "O" rings under cold conditions. A conservative interpretation of the effects of cold on the "O" rings would have led to the launch being delayed.
- Management should accept the more conservative advice of engineers involved in the evaluation of Category #1 systems behavior.
- It is not clear that management had sufficient engineering appreciation of the design limits of the SFB joints. These joints have duel functions—sealing the hot gases from the SFB and holding the entire SFB stack together during launch. It is not clear that NASA personnel understood the dynamics of the SFB stack during launch conditions of firing the SFB in strong winds, leading to SFB "wriggle" or oscillation, leading to opening and closing of the gaps in the joints. The "O" ring gaps would normally be sealed if the temperature was warm at launch time!

## 8.2.9 Tenerife Accident

- State air safety organizations need to prepare plans for the emergency dispersal of groups of aircraft to small airfields. These fields should set up rules for safe control to avoid Tenerife-type accidents between aircraft on the ground. These safeguards may be more necessary in the time of ISIS attacks.
- Communication quality between towers and aircraft should be regularly checked.

- Although standard language is used, it needs to be checked for communications between non-English-speaking persons to ensure tower personnel can give clear instructions.
- Tower personnel need to be given instructions in the management of safety situations, since in emergency conditions they are likely to be in control. When nonstandard ground maneuvers are being carried out, additional personnel should be employed to coordinate operations with the tower.

### 8.2.10  NPP Containment Sump Blockage

- Management needs to be aware of issues that could arise and present unknown safety problems and act on them.
- Interactions between materials and systems can lead to unsuspected effects, for instance, insulation falling to the sump or seaweed in cooling water intakes.
- Operators need to be informed of these types of problems and where the instruments and controls are positioned for them to be able to take appropriate actions.

### 8.2.11  Fuel Cleaning Accident

- Management needs to be aware of the risk of used-fuel operations, such as cleaning, because of the heat still being generated. They need to be aware of what problems could arise and take appropriate action to closely monitor the process.
- Management is always responsible for design and operational procedures, since they are responsible for the plant.
- One should run tests to see if design operates properly, before turning it over to operations personnel.
- An analysis of the operation should be carried out to see what could go wrong and what the result might be, especially if one is dealing with the removal of decay heat, as well as cleaning the fuel surfaces.

### 8.2.12  Replacement of Steam Generators

- There is always a risk for an operating plant to replace known components with unknown units—problems with choice of materials and design features. The conservative approach would be to replace with like or make small changes.
- NPP steam generators are complicated pieces of equipment; changes in design need to be reviewed by experts to try and ensure that they do not suffer from increased hydraulic forces.
- Management needs to be advised about the issues facing steam generator (SG) replacement. The wrong decision led to SGs at San Onofre failing catastrophically, leading to the shutdown of the station many years prior to its expected lifetime.

## 8.2.13   IMPACT OF MANAGEMENT, NU OPERATIONS

- There is always a risk for an operating plant in replacing the management with one of a radically different philosophy, unless there appears to be no alternative.
- The role of the board of directors is to review the philosophy of the new managers to ensure that their objectives are backed up by experience operating under these new rules.
- The board of directors represents the interests of the shareholders and needs to ensure that the management team is not only producing a profit, but also running the organization in a manner that ensures that the plants are safely and economically operated.
- The board is chosen to ensure management is running the operations well; this means that the employees' opinions should be considered in deciding if this is the case.
- Insights into the quality of operations can be gained by consulting with the regulator (Nuclear Regulatory Commission) and the industry operations groups, such as Institute of Nuclear Power Operations (INPO) or World Association for Nuclear Operations (WANO) or maybe the Institute of Atomic Energy Agency (IAEA).

## 8.3   SUMMARY

The review of the accidents has revealed a number of insights from the accidents and they have been listed previously. First and foremost, they indicate that organizations are not totally prepared for accidents, both major and minor. The indication seems to be that management can become distracted from the important job of avoiding accidents that could lead to the demise of the company/organization. One would think that this is the most important job for the management, but of course things like making a profit or maintaining or increasing shareholder value are important and a manager's performance is ranked on these things. However, the last thing that crosses the shareholders' minds is that the survival of the company is at risk. The shareholders are mostly interested in how much value is being added to the shares. In high-risk industries, however, it should never be far from management's thinking how to ensure survival, and how to minimize the risks associated with operation while at the same time enhancing the efficiency of the operations. This is an interesting balancing act and occasionally slips occur. It is hoped that the approach taken in the book helps management reduce the risks of operation by helping to support good decision-making.

# 9 Role of Regulation in Industrial Operations

## 9.1 INTRODUCTION

The regulation of potentially dangerous commercial activities such as nuclear power has been a part of the legal process for some time. The government should, as part of its activities, help protect the rights of individuals and minimize the effects of unsafe acts on the health and safety of persons. These activities of governments are spread over various agencies and each agency focuses on a specific industry, for example, the focus of the Nuclear Regulatory Commission (NRC) is on that part of the electric power utility business using nuclear power to generate electricity. It is not the intent of the authors to discuss extensively controls, regulations, and the development of such regulators for a range of industries. The focus here will be on the NRC and some of the functions it performs, which are helpful in reducing risk of accidents.

It should be pointed out that not all regulations are carried out with the same intensity as applied by the NRC. In fact, the NRC has been upheld as an example of an effective regulator, and other industries' regulators should be improved. For example, during the oil spill in the Gulf of Mexico, although the accident consequence could have been terminated if the blowout preventer had been more carefully maintained, the Mineral Management Service (MMS) regulator seemed to be more interested in leasing oil zones than in the safety of operations (see Section 7.6.1). In fact, the name and function of MMS has been changed over to the Bureau of Ocean Energy Management, which perhaps will perform a better service than the old MMS!

The United States is not the only country to have regulators; other countries have their own, such as the Institut de radioprotection et de sûreté nucléaire (IRSN) in France and the Health and Safety Executive (HSE) in the United Kingdom. The scope of the HSE is much wider than that of the NRC. The HSE uses a different approach to safety regulation and is based upon the concept of the "safety case." "The safety case is a structured argument for why the system is safe"; this definition is given by Leveson (2011b). Leveson has an interesting paper questioning the validity of the safety case. She also states in her conclusions that "worst case" should be considered, not just "design-basis accidents," and that all factors should be considered including management structure and decision-making! Her view confirms our view of the top-down aspect of consequential decision-making applied in all industries.

Outside the bounds of the country, there are other organizations in the business of nuclear safety, like the Institute of Atomic Energy Agency (IAEA), that are involved in both spreading knowledge of atomic energy applications and trying to ensure that the applications are safe.

The United States Atomic Energy Commission (USAEC) was set up to develop the nascent nuclear power industry (Atoms for Peace program under Eisenhower) and at the same time regulate the activities of the industry. Later, this arrangement was changed and the regulatory portion became the NRC and the other function, to help grow the industry, was subsumed by the Department of Energy (DOE).

## 9.2   REGULATION PROCESS

Responsibility for operating power plants lies with the nuclear power plant (NPP) owners. In the United States the regulator, NRC, has been set up by Congress to play an important part in trying to ensure that NPPs are operated safely and act to protect the public. The process is set up in such a manner that NRC promulgates rules and regulations that the operating utilities should follow. The NRC positions inspectors to reside at the NPPs to observe whether or not the utility follows the rules. The NRC is not empowered to cause the management to be changed or redirect how the company is operated. However, if the company fails to take action and follow rules, then the NRC can cancel their license to operate. This was the case with Northeast Utilities (NU) (see Section 7.9.4).

The NRC by its actions cannot prevent or stop an accident occurring; it can only set up a process to direct the utility into a position whereby the utility takes the appropriate steps to ensure that the plant is operated safely. If an accident occurs, it is analyzed by the NRC and new rules may be generated to help prevent a similar accident occurring. This is a very reactive process, of course, and the hope is by having NRC inspectors at the site, they will see situations developing, advise their management to intervene, and thus prevent an accident. The review of accidents and incidents indicates that the chances of the inspectors detecting situations that can lead to accidents are limited; often utility personnel are the ones that disclose to the inspectors the near-accident situations.

Other influential bodies in the picture are the Institute of Nuclear Plant Operations (INPO) and the World Association of Nuclear Operations (WANO). INPO is a US-based organization working with US utilities. WANO acts outside the United States and covers many of the same fields as INPO; in fact, it often uses INPO personnel to further its effectiveness. INPO can help a utility, if asked, to improve its operations by assisting with training of utility personnel, performing reviews of utility operations, and being a vehicle to pass information on good practices from other NPP operations. INPO also records problems that have occurred at other utilities that might undermine operations at a utility's NPP and passes this information to the group of INPO utilities.

One factor that is not seen to be very influential in the operation of NPPs, but in fact is very important, is the cost of power on either the open market or as determined by the utility's public utility commission (PUC). The cost of power can have an effect on the attitude of management and how it runs the operation. It is easier to afford good people and take time to maintain equipment when one is not too concerned about profit margins. However, if the market squeezes the NPP operation, then it takes excellent management to handle financial issues and still run a safe plant. If the state is aware of the needs of NPPs to be viable from both a safety and financial viewpoint, then it can apply rules to help ensure the plant is operated safely and the utility is economically viable. However, if the NPP management is not good, even if lot of money is available, the plant could be still run in an unsafe manner. The rules that apply to management functions must take into account the needs of the utility as a whole. The viability of the company depends on the management processes and how effectively they are embodied in the operating rules. The top management, in particular, has the responsibility to operate in a manner that leads to the utility fulfilling its role as being both a safe operation and run in a financially prudent manner.

## 9.3  LESSONS FROM REVIEW OF NRC REPORTS

Accidents have had a measureable impact on high-risk organizations (HROs), but are the issues associated with poor management decision-making visible only on these occasions, or are they present at other times? Some lesser accidents were listed in Chapter 7. These indicated that poor operations could lead to accidents. Other accidents/incidents can occur and still have an influence on the industry (see Section 7.9). It should be pointed out that these are not the only ones to have occurred. Incidents occur all of the time, indicating that utility management needs to be aware of operational deficiencies that can occur at any time.

An important program run by the NRC is its Reactor Oversight Program (ROP). This program can reveal insights into the operation of NPPs. Many issues covered in ROP are associated utility management decisions discovered during power plants' operation. Section 7.9 covered some management-induced problems. These were related to the management of NU in the years 1986–1995, and were also discussed in MacAvoy and Rosenthal, 2005.

The intent in this section is not to cover all the years of NPP operation, but to cover some parts of the ROP record to reveal some of the details contained within ROP records. A number of incidents that have occurred are discussed here. Observations of the historical data are examined to see how the observations are related to various organizational functions depicted in Chapter 3 (see Figure 3.3). The purpose is to see if there is any support for the view that management decision-making is a more important contributor to generation of accidents or incidents than those caused by operators.

A section of the ROP database has been selected for examination. It is assumed that the basic characteristics of the data within the selected database are typical and that no one recent year is very different from any other one, so the conclusions drawn are much the same for one year as another. However, there is one proviso: the impact of NRC attention can cause a utility to modify its attitude, taking it from one class to another. Equally, deterioration can cause a utility to drop as a good performer. Overall one could expect the characteristics of the data to remain fairly constant. This has not been proven here. It is believed that there are outliers that do not confirm to this assumption. The industry organization INPO is also believed to have a smoothing effect on the data by reducing the numbers of outliers; this by virtue of its role in helping utilities with training and organizational improvements.

The data referred to here was obtained from the NRC web site, http://www.USNRC.gov. Entry was made to the ROP portion of the web site and the current set of data was accessed. The ROP database allows one to gain access to the data which covers US stations. One can see from the data that most of the stations (units) are run very successfully, but some stations did not have such a good record and did experience some problems. Most of the problems were seen to be fairly minor, but others were of greater significance. Even the most significant issues were not accidents, but rather incidents. The data does indicate that the management of some NPPs is not as good as the majority.

Table 9.1 shows the list of "problem plants," of which there were 19. The rest of the US NPPs (84 units) had acceptable performance and were placed in a category whereby they are inspected occasionally, rather than operating under tight enforcement. The assumption is, for these nominally well-run plants, management is operating fairly effectively and there is no reason to carry out any further investigations. This means that about 22% are under a varying degree of risk and are under some enhanced monitoring by the NRC. The experience with the NU indicates that it is possible to operate plants poorly and not get into an accident. It is the combination of not operating the plant correctly and then being exposed to an unexpected initiating event that can lead to a major accident. The state of not operating the plant safely could be called a quiescent state of operation, with the utility just waiting for an initiating event to come along that could lead to a major accident. Events that could then cause an accident in the case of these poorly operated plants are something like the arrival of a large winter storm, like *Sandy*, or the failure of an internal pipe leading to flooding of a unit. The NU Millstone plants in this state could have been in the path of a large storm traveling up the coast, which could then lead to a severe accident.

However, as the NU experience informed us, the deterioration in plant performance induced by management activities can lead to an increased unavailability of plant systems and the increased possibility of risk of a major event. The Tokyo Electric Power Company's experience leads to the conclusion that failure of the management to understand the risks associated with external initiating events and failing to act to minimize their effects can end up causing the near demise of the company and long-term problems for the country.

---

**TABLE 9.1**
**NRC ROP Action Matrix Summary**

**ROP Action Matrix Summary (Up to Date April 2, 2013)**

| Licensee Response Column | Regulatory Response Column | Degraded Cornerstone Column | Multiple/Repetitive Degraded Cornerstone Column | Unacceptable Performance Column |
|---|---|---|---|---|
| 84 NPP units | 15 NPP units | 3 NPP units | 1 NPP unit | 0 NPP units |

The ROP data reveals that some 20% of plants do have problems, which can be minor to more severe. Examination of the NRC site inspectors' reports can reveal different problems. The inspectors' reports are grouped into the following areas:

1. Initiating events
2. Mitigating systems
3. Barrier integrity
4. Emergency preparedness
5. Public radiation safety
6. Occupational radiation safety
7. Security

The first four are associated with reactor safety. The next two are associated with radiation safety and the last with safeguards. The NRC inspectors in their reviews of incidents place the incidents into one of these groups. All of these groups are associated with internal events and do not cover the topic of external events, like earthquakes, etc. In other words, these events result from the activities or lack of activities of the NPP organization personnel.

Seen from the organizational viewpoint (see VSM Figure 3.3 in Chapter 3), these activities could be related to Systems 1 through 5. One can go to the NRC web site and see the reports associated with given plants. The reports point out which area (physical or functional) that the incidents are identified with and what the NRC thinks that is the cause of the incident. In these public reports, no person is identified as the cause of an incident. Also, the area covered under "security" is not covered in the data given on the web site, for obvious reasons, since it might give key information about security weaknesses at the plant.

Covered here is one particular plant, Robinson 2, for the four quarters of 2012. Figure 9.1 covers information on the significant inspection findings for the four quarters.

The figure shows that there are some G (Green) areas for events that occurred in 2012 that were reactor safety issues associated with initiating events, mitigating systems, and a barrier integrity event.

To illustrate these reports, an example is given as follows:

**Significance:** Jun 30, 2012
Identified by: Self-revealing
Item type: FIN (finding): **Lack of preventive maintenance on feedwater control switch results in an automatic reactor trip**
A self-revealing Green finding was identified when the licensee failed to establish adequate preventative maintenance for equipment associated with the feedwater control systems. Specifically, the licensee's inappropriate classification of the feedwater flow loop selector switch as a "run-to-failure" component permitted the switch to remain in service, without preventative maintenance, until its failure on March 28, 2012, which resulted in a feedwater transient and reactor trip. Corrective actions included the replacement of the failed switch and future replacement of seven

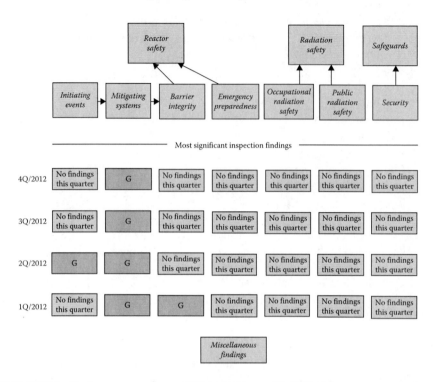

**FIGURE 9.1**   Performance summary 2013 for Robinson #2, 2012. (From www.nrc.gov.)

additional switches that were deemed to be at risk for a similar failure. This issue has been entered into the corrective action program (CAP) as Nuclear Condition Report (NCR) #527203.

The licensee's inappropriate classification of plant equipment in accordance with ADM-NGGC-0107 Rev. 1, Equipment Reliability Process Guideline, which permitted feed flow selector switch 1/FM-488B to remain in service, without preventative maintenance, until failure was a performance deficiency. This finding was determined not to be a violation of NRC requirements. The finding was more than minor as it was associated with the initiating events' cornerstone attribute of equipment performance, and it affected the associated cornerstone objective to limit the likelihood of those events that upset plant stability and challenge critical safety functions during shutdown as well as power operations. Specifically, the performance deficiency caused an automatic reactor trip from 55% power operations on March 28, 2012. The finding was determined to be of very low safety significance (Green) because the finding did not contribute to both the likelihood of a reactor trip and the likelihood that mitigating equipment or functions would not be available. The performance deficiency had a cross-cutting aspect of evaluation of identified problems in the area of problem identification and resolution, because the licensee failed to thoroughly evaluate the events in 2010 and 2008 such that the resolutions addressed the causes and extent of conditions as necessary [P.1(c)] (Section 1R12).

### 9.3.1 COMMENT ON REPORT

There are several points to be made from this inspection report. The incident was a reactor trip caused by a feedwater transient induced by the failure of the feedwater selection switch, so this was an initiating event, which could lead to a reactor safety incident. The maintenance personnel failed to categorize the switch correctly and placed it in a run-to-fail rather than in the maintenance prevention category. Apparently, there were other occasions when the maintenance personnel failed to do this task correctly.

The NRC does not further analyze the situation and identify whose responsibility it might be. Seen from a viable systems model organizational viewpoint, clearly the control-room operators were not responsible; they just responded to the event and did take the appropriate action. Maintenance staff was not directly the cause, since it would maintain the equipment as instructed. The ultimate responsibility rests with the chief nuclear officer (CNO) and his or her staff supported by the probabilistic risk assessment group looking at the effects of both equipment and personnel actions on the safety of the plant. The plant engineering group under the plant engineering manager should also be involved to help other personnel understand how the plant operates. The good thing here is that this was an incident identified by the plant personnel.

Taking this report together with Figure 3.3, we can see the involvement of S1 (maintenance supervisors), S3 (plant manager), and S4 (CNO). The maintenance categories ought to have been established initially by the NPP designers, but this was a long time ago. The control and protection (C and P) design logic was done in 1967 (private information: Spurgin was the designer of C and P systems). It is the responsibility of S5 (CEO) to have had a review of requirements carried out every few years. It is the responsibility of the CNO (and staff) to review the safety systems to ensure that the plant is not exposed to a high risk of failure.

Although the main responsibility rests with the utility to ensure that the plant is safe and not exposed to undue risks which can impact the economics of operating NPPs, the NRC (or other regulators) can perform a useful function of investigating seemingly minor incidents, which can reduce the risk of larger accidents occurring. Additionally, experiences with similar plants having a variety of problems can aid the utility to review its operation to see if the same type of problem is present in their plant.

## 9.4 COMMENTS

The objective of this book is to promote effective decision-making by management of HROs, so how do the regulation process and regulators help? One could say that it is in the interest of organizations to self-regulate. Unfortunately, the organizations' objectives may not be those of society and therefore society's interests have to be upheld and the effective mechanism for so doing is the NRC or equivalent.

Figure 4.2 shows the relationship between the organization running the plant and external inputs that affect the manner that the plant functions. Clearly, some of the inputs are environmental disturbances—tsunamis, which are initiating events of accidents—and others are effective as stabilizers, which can help an organization override the possibility of an accident by their advice and experience. The NRC and INPO can fulfill this role with the involvement of the utility management.

It appears that if one accepts this role of the NRC and finds ways to work effectively with it, things can go smoothly. This is not to suggest that the NRC or any regulator is without blemishes, but the best way to operate is to acknowledge the regulators' role and work with them. If you think that they are wrong, then discuss your logic with them and try to resolve the issue.

Management's role should be trying to ensure that the plant is operated as efficiently and safely as possible. Usually, regulators like the NRC have research capabilities and it should be possible to tap into their research work and learn from it so that mistakes made by others are not repeated. Also, the NRC has inspectors stationed at the plants; these are extra eyes, persons with questioning attitudes and capable of detecting problems that should be attended to, and they can supply warnings.

Situations in which the attitude of the managers affects how the plant is operated can lead to problems with the NRC (or regulator). Clearly, the experience with NU (see Section 7.9.4), was a case in which the management ignored the advice from the NRC and INPO for some time. Even an approach to the board of trustees was of no avail. At the last moment, before the NRC closed down the plants, the chairwoman seemed to realize that the NRC was serious and tried to rescue the situation—too late! On review, the NRC seemed to be very slow in exercising action and gave NU management and the board many opportunities to act to correct the situation. This situation reinforces the idea that the NRC and INPO are really there to try to get things to work for the good of the industry and public.

This experience drives home the point that selection of management personnel is critical. Organizations need to focus on management's critical decision-making skills, which should include the capabilities of learning from experiences and accepting advice from others. In the NU case, the advice was coming from outside the organization—the NRC and INPO. However, in the case of NASA (*Challenger* Shuttle), the advice was inside the organization, from an engineer (see Section 7.8.1). In both cases, the advice was not taken. The NRC eventually revoked the operating licenses for NU and in the case of *Challenger*, it blew up.

# 10 Integration of Tools Related to Decision-Making

## 10.1 INTRODUCTION

The objective of this chapter is to describe how the various tools covered earlier are linked to each other and their value in the whole scheme of improving managerial decision-making. It is clear that managers need help with the decision-making process as judged by their performance in dealing with accidents and other difficult situations (see Chapter 7). Management is seen needing help understanding the physical, safety, and economic conditions surrounding approaching critical situations. Examination of a number of the accidents/incidents has indicated the effect of managerial detachment on the effectiveness of plant operations in preparing and responding to accidents/incidents.

Operations can operate successfully without the involvement of the CEO/CFO. However, the organization needs the involvement of managers to ensure that the organization is pointed in the correct direction. Of course, one realizes that organizations need managers; an examination of the Beer model of an organization points very much to the need of cognitive processing to be able to lead an organization. Equally, an organization needs personnel to organize and take actions, as required. Ultimately, if the underlying idealism and philosophy is good, the organization should prosper; however, if the CEO and CFO do not have a good philosophic approach to dealing with the organization then control of the organization will go awry; this can be seen in the operations of Northeast Utilities (see Chapter 7).

Managers of organizations need more than a good philosophy; they need knowledge, experience, and understanding of how to lead an organization. Leadership comes from being trained properly and having confidence in their decisions. The methods of the nuclear Navy, as developed by Adm. Rickover, is based upon training and knowledge, but also training in the business of decision-making. His approach (see Appendix) was to develop the required decision-making skills by exposing submarine crews to more and more difficult problems. The training, in part, was built around using a land-based real nuclear propulsion nuclear reactor in the training process to lend reality to enhance the pressure felt by the "students."

Doing the right thing during a high-stress condition is not a gift that only a few are born with. It is an ability that some can learn better than others, but they have to be exposed to situations so they can better develop the required skills. Junior to senior members are tested, so they can gradually build up their decision-making skills. These skills range from running equipment to operating the whole submarine operation. Developing these skills helps in being able to terminate accidents and enhance the recovery process.

It is the integration of the elements covered earlier, plus the education in their use, that forms our view of good management decision-making and how it can help to reduce the probability of poor decisions and their consequences. The process rests on an approach similar to that of Adm. Rickover. There are a number of tools that were not available to Adm. Rickover that can enhance the ability of managers to understand the dynamics of organizations, the behavioral characteristics of personnel, and the training of managers to understand the fundamental characteristics of systems under various domains, so that they can be controlled. One tool that has been used effectively is simulation of the plant. However, one must understand that simulation techniques can have different uses and they are not only limited to simulation of plant equipment for training operators. For example, chess could be considered a tool to train persons in the concepts of tactics and strategy and is a simulation of old military operations.

Often, it is the poor decisions made by management that lead to accidents and to the failure to position the organization to recover from an accident (see the Fukushima tsunami accident).

Many of the safety techniques that have arisen in the nuclear industry have done so because of considerations related to public health and safety. We believe that while this is good, it is insufficient and there needs to be a consistent policy addressing how all persons are trained. Much attention has been given to operators; unfortunately the people who lead companies seem to receive much less training than operators, relative to the tasks that they perform. This seems counter to what it should be!

When does an operator working in the control room decide on the design of the safety displays, manning schedules, plant equipment design improvements, etc.? These people can make suggestions, but they do not make decisions, especially if money is involved. Even when the possibility of failure of the project is involved, the management makes key decisions, and even forces a decision, for example, the *Challenger* explosion in 1986 (see Chapter 7).

History has revealed that the safety aspect of nuclear power has been exaggerated and that the safety of many nonnuclear industry activities have been underestimated. However, it should be mentioned that over time safety issues have been tackled strongly in the nuclear industry. A key safety component to protect the public is the containment. The requirement for nuclear plants to have a containment was taken early in the development of nuclear power in the United States.

The use of a containment in the design of power plants has shifted issues from safety to economics. One notices that for later-design nuclear power projects (NPPs), principally with containment structures, the number of deaths is estimated to be reduced. If NPPs are operated correctly, the public should not be at risk (see Chapter 7). However, if an accident occurs and the release of radioactive material is controlled, there might be core damage resulting in a core melt. The short-term and long-term costs are such that the organization operating the plant may go bankrupt. The function of the management is to understand the characteristics of NPPs sufficiently well to prevent accidents, especially those that end up melting the core.

Many of nonnuclear industries have had accidents that are much more severe than nuclear accidents. Many of the nonnuclear accidents could have been avoided quite easily; for example, in the Bhopal chemical plant in India in April 1988, some 200,000 people were killed. If the site manager and remote managers in the United States had understood the dynamic of the plant and fundamentally known how to control the processes, that accident might have been avoided. Simply put, the managers needed to better understand the risks and take precautions! There are other examples to support this view. Developments in the nuclear industry have evolved over the last 45 years to reduce the probability of accidents. It is suggested that the same should have occurred in the nonnuclear industries, like oil and gas, mining, railways, chemical industries, etc.

Some of the techniques were developed from the consideration of the nuclear industry in its need to pursue safety. As the apparent safety of nuclear plants has evolved, concern over the economic effect of accidents has increased. One can appreciate this from the study of the Three Mile Island #2 (TMI #2) and Fukushima accidents. TMI #2 accident did not release significant radioactive materials and no one was killed in the short term. However, the core melted and the plant was shut down permanently. This led to the following costs: loss of plant, need for replacement power, the dismantling of the plant, and safely storing components and fuel away from the biosphere.

The items that we wish to cover briefly, and how they can be linked to help management decision-making, are as follows:

1. Beer's cybernetic model of organizations (Chapter 3)
2. Ashby's law of requisite variety (Chapter 4)
3. Probabilistic risk assessment methods (Chapter 5)
4. Rasmussen's skill-, rule-, and knowledge-based human behavior models (Chapter 6)
5. Case studies of accidents for different industries (Chapter 7)
6. Training methods and role of advisors (Chapter 8)
7. Simulation of processes and its value (Chapter 9)

## 10.2   INTEGRATION AND ROLES OF EACH ELEMENT

As mentioned earlier, all of the elements mentioned have a role to play in the approach to improve the decision-making skills of management. The central purpose/function of each element and its relationship to the other elements are covered.

### 10.2.1   BEER'S CYBERNETIC MODEL

Beer's cybernetic model is a key element, since it provides the structure upon which the understanding of an organization's operations rests. It depicts the functional relationships between the various organizational elements: the top management, support managers and staff, and the operational personnel (operators). The Beer model is dynamic, in that instructions, guidance, and status information on operations/actions are flowing and connect the various elements. However, one has to consider Beer's organization model relative to the system that it functions to control, so one can see that the organization acts as an intelligent control system. The Beer model is discussed in detail in Chapter 3.

Figure 10.1 depicts the organization in the form of a Beer model controlling a nuclear power plant. The plant components are automatically controlled and protected by a series of control and protection systems designed to keep the operating parameters of the power plant within design limits, which should be safe. The automatic systems are backed up by control-room and other operators reporting to the plant management. The power plant is operated to provide electric power to the national grid. The grid requirements are met by the control-room personnel by operating the plant within limits. Other personnel are involved in ensuring the plant equipment is maintained in good working condition.

As depicted in the figure, other organizations, such as the nuclear regulator, monitor the plant operations to ensure that the management is operating within its license agreement. Also seen in the figure are external and internal disturbances. The external disturbances are things such as earthquakes and storm-induced floods, while internal disturbances are things like internal floods caused by pipe-breaks, etc.

An operating system, which is being controlled by an organization, could be a nuclear plant, a refinery, or some other enterprise. When we talk about a system, it is the understanding of the dynamic behavior of the system in detail that is needed by the management team. This connects one to the use of Ashby's law, if one needs to understand how to control the system. One also needs to know how both external events (such as floods, earthquakes, etc.) and internal events (such as internal floods and fires) can change the behavior of the system. The organization is also influenced by laws and rules enacted by governments. These laws and rules can also modify the decisions and actions taken by the organization.

### 10.2.2  ASHBY'S LAW OF REQUISITE VARIETY

The essence of Ashby's law in this case is the awareness of the key role of the top manager (controller) in understanding the characteristics of the process and its environment, to be able to control the process effectively. The failure of the decision-maker to understand the processes can lead to the failure to adequately control the process, both before, during, and after an accident/incident.

### 10.2.3  PROBABILISTIC RISK ASSESSMENT STUDIES

Probabilistic risk assessment (PRA) can be used to make assessments of risk of management decisions and for the examination of plant risk for short-term equipment changes made for operational reasons, that is, the removal of a safety system from service for maintenance purposes. For this latter case, one considers the integrated plant risk profile, so the organization can decide to increase the monitoring of the plant to effectively reduce the impact of a coincidental accident initiator.

The components of a PRA are a range of accident initiators: the probability of important equipment failures, coupled with the probability of a key human action. The PRA generates a complete risk profile of an operating plant, given the capability

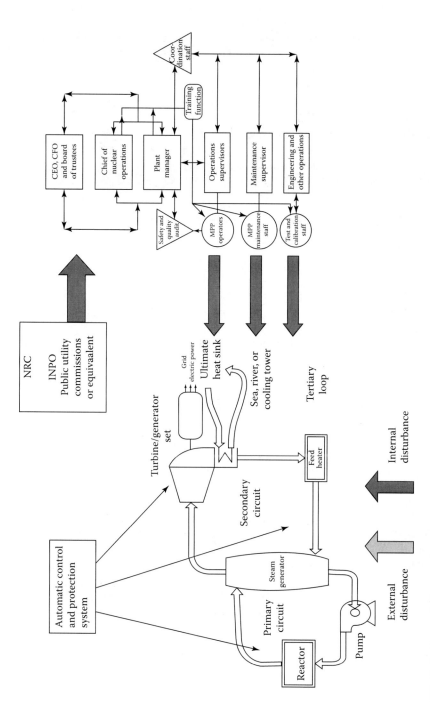

**FIGURE 10.1** Integral diagram showing combination of a system, organizational model, and other influences.

of analysts to predict the various probabilities: accident initiators, equipment failure, and human errors. It takes very experienced analysts to predict these probabilities; often the probability estimates include a measure of uncertainty, so one gets/arrives at a risk profile including uncertainty. Clearly, this gives managers a tool to make judgments about what actions to take; however, the decision-maker has to have faith in the probabilities estimated by the analysts. This is another of way one can see the value of Ashby's law—the manager has to have a deep knowledge of the workings of a PRA system and take into account the uncertainties when making decisions. Earlier, the training of staff and managers were discussed; this is one element in the package in which management has to be trained! The PRA is a tool that has known limitations, but it can be something that the manager takes into consideration in deciding a course of action.

### 10.2.4 Rasmussen's Human Behavior Types

Rasmussen worked in the Riso laboratory in Denmark on a number of topics, including human factors and psychology. One of his contributions was the concept that human behavior could be categorized and divided into three categories: skill-based, rule-based, and knowledge-based. Of course, the three categories are really a continuum, but it is useful to break them into three, namely: skill-based, rule-based, and knowledge-based behavior.

Skill-based behavior is when a person engaged in a repetitive task becomes so skilled that he or she does not need to refer to advice or help from any person or documentation. A simple example of this would be the skills of a woodworker in using a chisel.

Rule-based behavior is when a person is skilled in carrying out a task by following a procedure step by step. An example of this process is the actions of an operator following an emergency procedure when responding to an accident on a nuclear plant. Clearly, the design of the procedure has to help the operator achieve the correct series of actions. The operator should have been trained in the understanding of the procedure and its applicability to the accident. Part of the procedure package is a component to lead the operator to the precise procedure to be used to combat a specific accident.

Knowledge-based behavior is when a person uses his or her knowledge of a system or process to tackle an unknown accident. In performing the task he or she is led only by understanding and not by some procedure. A person competent in the technology could be capable of taking the correct actions to terminate or mitigate the effects of an accident, but it is a question of how long it takes him or her to reach that state. In an academic situation, time may not be an issue, but in an accident situation, time is important and the consequences can become much worse as the failure to take an early action may lead to even more complex situations.

### 10.2.5 Case Studies of Accident for Different Industries

Accident analysis is very useful in a couple of different ways. Analyses indicate to persons that accidents can happen, even in the case of nominally well-run organizations, and also the accident sequence can indicate that the organization needs to understand that accidents are often multifaceted and interactive; so that dealing with an accident, one needs to consider the different pathways that the accident could take. Often, in the design of procedures to combat the severe accident progression, the interactive effects of, say, tsunamis and earthquakes late in the accident may not be considered but organizations need to consider these situations. An example of this type of difficulty is the Fukushima accident. In this accident, the recovery actions of the station personnel were affected post-accident by the ripple effects of the earthquake, which impacted the steps being taken by the crew under difficult conditions. Crews were setting up auxiliary electric feeds and fluid lines to inject water to cool the core, when smaller earthquake-released debris led to these emergency facilities being damaged and delaying actions.

### 10.2.6  TRAINING METHODS AND ROLE OF ADVISORS

So what is the role that training fills? The decision-maker may be appointed to a job for a number of reasons; in fact, it has been observed that some persons selected are without a deep background in the field. So the function of training is to prepare persons for this task. As the military has shown, privates are not promoted to be generals! In the military it is a requirement to train novitiates by a graded process of increased understanding of war tactics and strategy along with testing to see how the soldier is processing. The conditions selected for decision-making under stress conditions for the novitiate are less difficult than for higher ranks such as captains. The military tries to set up conditions to get the person to fail, so that he or she gets to appreciate the issues associated with the operations and learn. Unfortunately, in private industry, training is not as rigorous and persons can be chosen for leadership positions whose background is very limited and their capabilities are untested under difficult stress conditions. In the nuclear industry, the training requirements are more intense for operators than they are for top managers, so some organizations do like their managers to have been trained as operators. However, there does not seem to be a requirement for continuous training, like that required of operators. Operators are cycled continuously through simulator training for a week about once a month; they also have classroom training.

### 10.2.6.1  Use of Advisors

The set of tools that a manager uses to form a decision includes his or her training and experience, including accident analysis, and also uses advice generated from the use of the PRA system. In addition, a manager can also rely on members of his or her management team for advice. It is possible that their knowledge and experience in certain matters can be greater than the managers'.

A review of utility organizations shows that persons of deep understanding in nuclear operations are being promoted to the position of advising the CEO on nuclear matters. These persons are identified as the Chief of Nuclear Officers (CNO) and supposed to be both knowledgeable and experienced in nuclear-related matters. This appears to be a progressive move to fill a deep need.

### 10.2.7 Simulation of Processes and Its Value

Simulation is the process of representing systems dynamically by sets of mathematical equations. So, for instance, one can represent the behavior of a nuclear plant by a set of mathematical equations for the reactor, the steam generator, electric generators, steam turbines, and ancillary equipment. Simulators are used to train operators, when coupled to an interface duplicating the control and instrument systems of actual units. Other uses of simulators are to study the behavior of the plant during different accidents. For accidents, the mathematical representations may differ from those used for the purposes of training operators.

So the use of simulation models may be different. Typically, full-scope simulations are used in training operators and occasionally helping analysts understand accident progressions. The so-called accident models are to help analysts and designers of accident procedures. These procedures are used by operators to enable them to respond to complicated multiphase accidents. This brings us back to Rasmussen and his behavioral models.

### 10.2.8 Summary

As mentioned previously, this chapter was to cover the various tools and their relationship and how they could be used together by management to improve their decision-making, particularly with respect to accidents. Here are some notes to cover the relevance of tools for managerial use and other tools.

#### 10.2.8.1 Organizational Dynamic Model

Beer's organizational structure provides a dynamic model of an organization and can support a manager in understanding how an organization operates and the interdependence of all of the components from the CEO, CFO, lower managers and operators, and maintenance personnel. The structure also depicts the communications within the organizations without which the organization is unlikely to function correctly.

Just knowing about the organization is not sufficient, one needs to know about the dynamics of the system under all operating modes, including normal operation to accident conditions. Especially, one needs to know the effects of accidents on the dynamics and responses of the system. This is what an understanding of Ashby's law gives one and this is why it is included in the training package for management. It does not hurt to start introducing this to junior personnel, much after the principles of Adm. Rickover. Since management works to accomplish aims through people, it is well for management to understand the value and limitations of human behavior. Insights into human behavior in the working domain are obtained by understanding Rasmussen's behavior classification. Management can then see that personnel are assisted by training and procedures appropriate to the tasks they have to undertake. Also, when the requirement of the task puts one into a knowledge-based condition, then management should appreciate the need to allow staff sufficient time to solve the problem.

## 10.2.8.2 Ashby's Law

An understanding of Ashby's law is needed to be able to appreciate the variety of complex systems in order to be able to effectively control them. A failure to understand the variety of a system, in a given situation, means that one may not be able to control the system and meet the organization's objectives.

One also needs to understand how accidents can affect the characteristics of systems, again for the same purpose. Managers need to be aware of the possibility that their view of system performance can influence their actions relative to effective control of the system.

## 10.2.8.3 Probabilistic Risk Assessment

One of the manager's tools is a complete risk assessment of the plant. This approach could include all manner of plants, including nuclear, oil refineries, aerospace endeavors, etc. What the PRA does is lay out the relationship between external and internal disturbances, maintenance and operations human activities, management decisions, and consequences.

Management could use this information to decide on plans to improve plant-related weaknesses and in so doing reduce the probability of failure during accident situations. However, the class of initiators falls into different groups; some are related to the operation of the equipment and its age and others are associated with natural events, whose frequency is unpredictable. Equipment failure initiators can be improved by deciding on an inspection regime to control the risk and corresponding consequences. For example, the main steam line from a steam generator to main turbine should be ultrasonically inspected often for cracks or corrosion-wall-thinning, to ensure that the random line failure probability is heavily reduced and therefore the pipe break is highly unlikely to occur and cause an accident.

The estimated probabilities associated with given consequences should be treated with care. Low probabilities do not mean that the event will not occur soon. The message from the Fukushima tsunami accident is that even events with a 1 in 1,000 years' probability can occur the next day or next week! In the case of TEPCO, they decided to "hope" that the large tsunami would not occur, whereas the manager of the Onagawa NPP decided to build a bigger wall and the effects of the tsunami were minimal versus the large damage at Fukushima.

## 10.2.8.4 Rasmussen's Human Behavioral Models

Rasmussen's behavioral models tell us of the limitations of people to achieve given functions and their need for additional support to operate effectively. In the case of operators responding to an accident, their need is for procedures, which they have learned, to help them respond quickly and accurately. Rasmussen tells managers that to achieve good results in response to accidents, they need to ensure their operators receive training in procedures and that the procedures need to be designed and technically tested to ensure they do what is required.

The process of designing the procedures is one that falls into the field of knowledge-based behavior. Managers need to be aware of the basic assumptions in the procedures and agree to them, since that is central to their responsibility as far as safety and economics are concerned.

### 10.2.8.5   Accident Case Studies

The function of studying accident cases is to enhance management's understanding of how a lack of understanding of the plant processes can lead to accidents. A failure to consider the possibility of an accident can lead to the loss of the company, large loss of life and the houses of the public, and even the loss of agriculture due to contamination.

This ties management responsibilities to an understanding of not only the dynamics of the organization but also the vulnerabilities of the plant to the local environmental effects, which need to be factored into their decisions. Even if the risk is small, it is better for management to consider the impact of these events.

By examining accidents in other industrial fields, management can experience the fact that even well-prepared organizations can fall into the trap of not being totally prepared and make decisions that lead to accidents. Sometimes the accidents are caused by poor decisions and other times by not making decisions. Examination of the case histories in Chapter 7 reveals a number of different types of management decisions. By being aware of these cases, they can check their upcoming decisions against these different causes and then try to make the correct decision.

### 10.2.8.6   Training Methods

It is seen as imperative to apply continuous training for operators at nuclear power plants. It does seem that a modified version of this approach should be carried out for other high-risk organizations or for those who make decision at the managerial level. Basically the idea for such continuous training, both in technology and exposure to simulated accidents, is related to an understanding of both the behavior characteristic of humans (Rasmussen) and the need to keep those skills up to a given level. The training methods should match the needs of the managers and the operators. A modified training approach was developed as the response to the TMI #2 accident.

This approach could just as well be applied to other high-risk industries, since the need for operators to respond accurately and quickly is much the same as for nuclear. The result of failing to have a comprehensive training approach can lead to the same exposure as far as the company is concerned: large loss of life, destruction of the plant, and the pollution of the neighboring area. Of course, the distribution of the effects may be different—more or fewer deaths, etc. So this underlines the case for training operators relative to the incipient risk.

There does not appear to be an awareness of the training needs for the rest of the staff and managers. Some managers are implicitly trained, because they served their time as operators. But often the senior staff and the members of the board of directors have next to no understanding of the appropriate technology. These personnel could be lawyers and financial persons, but they do represent the interests of the investors, which presumably covers a need to ensure the plant is run not only economically, but also safely. One should recognize the failure of the board of trustees in the case of Northeast Utilities to fulfill that obligation. Therefore, the members of the board of directors need to be trained in characteristics of the plant to ensure that they can perform their task of representing the investors!

The CEO and other members need to be very aware of the plant's technological characteristics so they can assess the operational risks associated with running the plant. This is different to the former approach of fixing things, if they break. These people need to be trained in the meaning of Ashby's law of requisite variety, so they can position themselves to request staff to examine the likelihood of various pieces of failing and leading to internal or external events.

In order to operate the plant successfully, managers need to ensure the operators are well trained and they, themselves, are equally prepared by being trained in the technology and understanding the role of operators and the maintainers of the plant equipment, including the safety equipment. Since it is difficult for top managers to be aware at all times of incipient accidents and their results, it is suggested that some of the support staff should be highly trained in accident awareness and their progression. These people should be capable of giving good advice to management. In the case of some US nuclear utilities, there is an appreciation of this need and they have instituted the role of the CNO. Sometimes, it is a question whether the CEO is capable of taking advice, especially if he or she has different ideology. The case of Northeast Utilities reveals that an ideology can so affect one's thinking, which led to the demise of the utility as an economic entity.

### 10.2.8.7  Simulation Processes

The value of simulation of processes can be shown to be useful in many ways. Plant simulations are used for training purposes. Here the simulation is a mathematical representation of the whole plant dynamics. A replica of the control room is constructed with realistic representations of the displays and controls. Signals to and from the interface drive the plant simulation. Disturbances, which affect the performance of the simulated plant, are selected by the training instructors during the training sessions, and the responses of the control-room crews are recorded for later analysis and discussion as part of the training processes.

Accident analysts also use simulations of the plant, at a different level of representation, to study the physical responses of reactor, steam generators, flow dynamics, etc., to disturbances. These simulations are very detailed and complex, whereas the training simulators make use of so-called lumped simulations of the reactor, steam generators, pumps, etc. Training simulators produce an adequate representation of a plant's transient behavior to disturbances, sufficient to allow the crews to understand how the actual plant will behave responding to given disturbances.

The training of station crews seems to more than adequate for the purpose; however, training of top managers to understand risk situations from a technical point of view does not seem to be adequate in most high-risk operations. The nuclear Navy under Adm. Rickover achieved a measure of success in dealing with this issue by training all submariners on real land-based reactor units.

The Navy personnel at various levels occupied their normal positions from maintenance to sub captain and engineering officer in these training environments. The submarine engine room to the central control area was not a simulation, but a real operating unit. Adm. Rickover was not totally in favor of the concept of a mathematical base simulation of the plant, but realized that doing what he did was not possible for utilities. The trained sub personnel could move from land to sea operations fairly smoothly, so the roles of captain and engineering officer received their training at the land-based equivalent of the actual submarine and became qualified decision-makers, or they would not progress in the Navy. Training in submarine tactics and strategy is separate issue.

Operation of a power plant or any large installation is of a different order of magnitude than a submarine and the complexity of the plant requires the management staff to be aware of not only the risk of operation, but also the economical requirements for the plant to be operated both safely and economically. Top managers often take graduate business classes to enhance their training in business/economics, but neither the topic of operational safety nor the study of accidents and their prevention is emphasized by undertaking training courses. Some managers do rise through the ranks and benefit by being trained as reactor operators. However, being trained as an operator makes one aware of the reactor-plant characteristics, but does not necessarily develop one's skills in decision-making under high-stress conditions. Remember, operators are trained to follow procedures so as to efficiently deal with accidents. In the manager's role, the requirement is to better understand the complexity of reactor and plant dynamics reacting to unknown disturbances, that is, operating in a more knowledge-based mode.

There seems to be an attitude that the safety aspects are attended to by following the guidance and rules of the regulatory authority. For all organizations, they are entirely responsible for their own salvation. The regulators can help, but if something bad happens, the regulators are not responsible and continue going on their merry way, whereas the affected company can end up in the dustbin of existence. Of course, they may be lucky and have the state come to help and cover their debts. But there is always a price to pay. Some industries, like the deep-sea drilling operations in the Gulf of Mexico, did not have an effective regulatory oversight authority at the time of the *Deepwater Horizon*/Macondo drilling operation (see Chapter 7). BP should have taken more care in the design of the oil cutoff valve (blowout preventer [BOP]). In our minds, we believe that they had a poor understanding of protection system design covering redundancy, diversity and separation, and periodic testing to ensure operability. These are standards in the nuclear industry that have been understood and tested many times. One thing did come up from this accident was the role of the US government. Despite the fact that the accident was the responsibility of BP, the responsibility of the US government is the protection of Americans from hazards including oil spills. If the government had acted quickly and accepted the help of other countries, the effects of the oil spill could have been significantly reduced. Society as a whole has the responsibility to help in these situations.

# 11 Use of Simulation for Different Operations

## 11.1 INTRODUCTION

Simulations and simulators are integral parts of industries' decision-making processes. The complexity of systems in industries has grown in recent years and the days that systems were simple to understand in and deal with have passed. This situation is compounded by the fact that the advance of complexity has intensified. Systems, which were decoupled, are now closely integrated and are more difficult to understand in how they respond to various stimuli. It appears that the only solution to this evolution is to make use of simulations to determine the dynamics of these systems, so that we can appreciate and learn how they can be operated. Simulations allow us to unbundle the components of systems and then test the parts. We can then integrate these responses in our minds to comprehend the totality of systems' behavior.

Simulators and simulations are similar, but not identical. Simulators are mainly packaged digital simulations developed for specific purposes and encapsulated with an interface for persons to communicate with the simulation. Simulators can be used for many purposes, such as training plant operators, airline pilots, gun crews, ship pilots, etc. These simulations are used to enhance the training of personnel in dealing with complex situations to help improve their judgments.

Simulations are dynamic computer programs that can replicate plant or any other dynamic systems that can be represented by mathematical equations. Persons can then use these simulations to predict transient changes in systems and processes. They are used to design control systems and to study accident progression, dynamics of flame propagation, movement of stars, etc.

## 11.2 SIMULATORS

The whole field of simulators and simulation has opened up over the last 80 or so years. The idea behind the simulator is to train persons. So simulation is something that takes the place of real systems, but reacts similarly, though usually without many of the disadvantages of the system itself! Simulators can be used to train persons to operate dangerous equipment, without exposing the trainees to risk while learning.

One of the first uses of simulators was for training pilots, the so-called Link trainer (Link, 1942). Many pilots were trained during World War II (WWII) using the "Blue Box" as part of their initial training experience. Although the Link trainer was limited, the possibility of the trainee getting killed or injured was reduced while being trained. It presented a step in the direction of helping the pilot learn some of the necessary skills: communications, exposure to the sensation of roll, turn, and dive. It saved governments time and money in helping to train pilots quickly, at a time when many pilots were needed in WWII.

Later, flight simulators were much improved beyond the Link trainer, but were still physically tied down. One improvement was to give the pilots something they said they needed to fly: the sensation of flying by the seat of one's pants. This was done by placing the cockpit/cabin on a hydraulic lift. The movement of controls in the cockpit was tied via mathematical equations based on computers to better represent the actual plane responses to aileron, elevator control, and rudder movements taken by the pilot using his controls. Data from the math-simulator was displayed on instrument devices in the cockpit, such as pitch, climb, roll, etc. The synthetic environment around the plane is displayed on video screens that replaced the cockpit windshields. The scene changes with respect to where the plane is "flying." So, one can see clouds, cities, mountains, airports, and runways. All of this gives the pilots a sense of reality. Using this method, the pilot's abilities to follow all of the requirements can be examined, without flying an actual plane. This results is a good training environment with huge savings in fuel and the use of an actual plane.

In the nuclear power industry, simulators are used to train operators, in conjunction with procedures, to combat accidents. Some of these could lead to the destruction of the power plant and possibly hazard the nearby population with radioactive fallout.

There were a few simulators early in the development of nuclear power, but following the Three Mile Island (TMI) #2 accident, the United States Nuclear Reactor Control (USNRC) brought in regulations to require utilities to install simulators (per each different type of plant design) for the training of control-room crews to control accidents and mitigate their effects. These nuclear power plant (NPP) simulators are very accurate representations of the displays and controls on boards of the actual control room. The design and layout of a simulator training center is an attempt to ensure that all of the aspects of operator situation awareness relative to a real control room are maintained. So all the displays and controls duplicate the actual units, as far as the operators are concerned. Behind the interface, the reactor and the rest of the power plant are represented by a set of mathematical equations. Initially the focus was trying to model the reactor behavior accurately with less emphasis on the rest of the plant. Over time, it was realized that the rest of the plant should be modelled to more correctly represent the dynamics of the other plant components. Currently, the simulators are reasonable models of the complete plant and provide the training environment to help the operators meet their training needs.

Training simulators have not only been used for control training purposes, but have been used in research studies related to how control operators respond to accidents and the kind of errors they make and whether they are typical of errors made during real accidents (Spurgin,1994). The results from these studies are included in probabilistic risk assessments (PSAs).

Adm. Rickover's training requirements for the Navy submarine crews were different. He used actual reactors as part of his approach, expressing that, in his mind, reactor-plant simulators were not good enough to perform the required training function. Later, he realized that though it was possible to build submarine reactors on land, building civil reactors for this purpose was not realistic. The US NPP industry is in the process of converting to digital equipment and eliminating analog devices, so the control-room interfaces are being changed to match this industry-wide change. Control rooms are moving from things like strip chart records, dials, and other indicators to computer-based displays. An earlier control room was in some ways a better design for operators in that it had a parallel layout of information, not serial. The current layouts are a series of computer screens presenting serial information, so the operators have to investigate the state of the plant to find the information they need, rather than having it displayed at all times.

## 11.3  SIMULATIONS

The discussion above gives an introduction to the whole field of simulation. Simulation techniques have been used extensively in a number of different fields. In fact, the authors have been involved in a number of them. Simulation of systems provides one with the capability to study how systems behave without having to construct anything. One can then determine things like stability, feasibility, needed changes in designs, etc.

A word should be said here about Ashby's law of requisite variety and its relationship to simulation. One of the authors was employed to design the control and protection systems for NPPs (Spurgin and Carstairs, 1967). This is exactly the issue that Ashby identified; one needs to understand the requisite variety of the systems and how they interact in any plant in order to be able to design the controls for the plant. In fact, one needs this understanding to control anything, even an economy. Of course, making a simulation of the economy is a large task. Forrester tried (Saeed, 2015)! But the issue is, how do you discover a system's variety? For simple systems it might be possible to do this mentally, but as the complexity of a system increases this become more difficult. It appears that the only approach is to use simulation of the systems and gain insight into the systems' behavior and their variety.

The prior control systems, applied to both nuclear and fossil plants, etc., were ad hoc designs. It was decided early in the design of commercial NPPs to study the whole nuclear plant in order to sort out how all the parts worked together, so that the unit would be safe. The design of the control system was such that response of the plant should be stable for all operating regimes. In fact, methods developed by Lyapunov were used to study the stability of the plant. Later, simulation methods were found more to be useful, and required about the same amount of effort! The utility of the simulation approach was that it could be used to check stability and also set the characteristics of the plant's responses to various disturbances. Essentially, the requisite variety was what was needed to be understood. It is not claimed that one was aware of Ashby's law at that time, but one can see the problem was solved implicitly by following Ashby by building a simulation of the complete plant; reactor, steam generators, pumps/valves, etc.

The simulation was based on first-principle models: conservation of mass, energy, etc. rather than derived models. So the plant model contained the variety necessary to enable the system to be controlled. Without having the simulation model, it would have been difficult to be sure that the control system was capable of controlling the plant throughout the working load range, 25%–100%. Under low load conditions, the plant and equipment behavior changes, that is, the *variety* changes. The control system design has then to match these changes in order to operate successfully.

Other simulations are created to cover the domain of power plant accidents. Again, these models are there to define the variety that can exist under different accident conditions. The objective is much the same—to be able to understand how the system behaves when affected by different stimuli (initiator events) and then determine the best way to control the accident progression. In these cases, once an accident is under way, does one have the means to terminate or mitigate the effects of the accident? One also has to address the question of all the different accident initiators. The solution is to find ways to prevent all of the accidents; for example, in the case of a Fukushima tsunami accident it would have been to build higher seawalls!

In addition, to understand accidents, one needs to examine the behavior of groups of components in the form of the complete plants. One then has to study the behavior of reactor cores, steam generators, turbines, etc., and then integrate them for specific investigations. The objective of the individual investigations is to study the dynamics of each of the units under adverse conditions, like the behavior of the fuel elements of a reactor under conditions of a loss of cooling, or for steam generators under full flow conditions to see if the tubes vibrate. These simulator models are very detailed with much attention to building combinations of mathematic elements, whereas the other simulations were more like lumped equations representing whole units, like complete heat exchanges.

So there are a range of simulation models used for different purposes for control investigations to model different accident sequences or to determine the detailed effects of different forces on given components.

## 11.4   FUTURE USES: DECISION-MAKING

The previous sections have dealt with simulators and simulations and some of their uses. These methods and techniques have been very helpful in training, in the design of control and protection systems, enhancing knowledge of how accidents progress, and the effect of transients on various components.

A field that needs to be expanded is how to teach decision-making to managers, so that the probability of poor decisions being made that lead to accidents and corresponding bad consequences—death and destruction—are reduced. We have seen in various chapters that some decisions made by managers have been poor. Reviews of the accidents have revealed that some thought, consideration of the circumstances, and advice from others could have prevented a number of the accidents and situations. In fact, it was the idea of the authors to discuss accidents so that readers could learn the lesson that most accidents can be avoided.

The concept of chess as a learning tool for generals in olden days was introduced for them to learn the art of tactics and strategy. The great idea behind this is that a master in the art of tactics and strategy battles against a novice. The master then uses his experience and knowledge to instruct the novice by playing the game. The master can continually upgrade his game. The novice is forced to keep trying and in the process improves his skills. In the old days, these skills could be reflected in actual warfare.

A number of military organizations take a similar approach in the teaching process, to upgrade the skills of the trainees. The recruits are given some initial training and then they get to operate in progressively more difficult terrain, with obstacles, aggressive opponents, ambushes, live ammunition, etc.

The approach taken by Adm. Rickover was a graded approach, with his officers being exposed to more difficult decisions taken under pressure to prepare them for the different roles, from junior engineer to executive officer. He also believed some would be less capable, so those failed. Not all persons are capable of being good decision-makers under stress, but it is necessary to select those who can succeed. Appointing persons because of some other criteria can lead to defeat in the field and the loss of power plants due to an accident (see Appendix for information about Adm. Rickover's management principles).

This is the kind of progressive intense learning approach that is needed to enhance the ability of decision-makers. It does not seem reasonable to organize something as intense as marine battle training in order to select a manger to operate a nuclear plant or something equivalent.

What could work is something like chess, but of course adapted to the science of today. The idea of a master guiding the learning process of the trainee seems useful and an advance on the usual apprentice process. The scope of the system could change to match the level of the trainee within the company/organization. The trainee's progression would match his or her decision-making role within the organization. The organization could have a training process going on over the years, becoming progressively more difficult for the individual to match the specific needs of the higher levels in the organization.

## 11.5  SUMMARY

This chapter covers the concepts of simulation and simulators and their usage to prepare the staff of organizations and constructors in the design of control systems. Simulators can be used for the preparation of managers in the matter of making decisions related to their business field. Simulators can be used to help in the training process of operators, from aircraft pilots to control-room operators of industrial plants to maintenance personnel.

The use of simulators can help plants, planes, etc. run more smoothly, more economically, and more safely through their part of training and design organizations. They should be considered valued components of an organization.

# 12 Training Approaches for Management

## 12.1 INTRODUCTION

Education is at the center of civilization's progress over the years. Education was exclusively for the elites of society, technological progress was slow, and life for the rest of society was a mean existence. Agriculture was the employment for the majority of people. It was not until the industrial revolution occurred that the new wealth generated spread to the lower classes, and along with this, persons could be involved in the intellectual activities of teaching and learning, because more money and time were available. This whole process caused industries to grow, and along with growth there was the need for trained persons to fill new job categories. This expansion process has continued and manual tasks have given way to more intellect-driven jobs, and with this more universities were started and technology has expanded exponentially.

The rush of technology development has tended to wash up some people on the banks of the river of technology. Yesterday's tools are just that—yesterday's. Who knows what a slide rule is—and, even more testing, how do you use it and what are the tasks you can solve with its aid? This technology has passed and is now confined to a museum. But what about the group of people who were trained in this technology? Have they become retrained or just drifted into the world of technical obsolescence?

The time scale of technical change is much shorter than the transit time in going from an entry task in a company to the top manager's job, so how do people keep pace with the changes? Or do they do what is stated previously—drift into a world of technical obsolescence? The job of a top manager is challenging enough—in effect looking into the future, as far as changes in economy, but there is also the possibility that an unexpected accident could occur for which he or she and the rest of the organization might be unprepared.

## 12.2 EDUCATION

Education, it has been said, is a lifetime experience, not a one-time experience. In order to keep up with technology one needs to be involved in updating one's knowledge. The requirement is not to just acquire more degrees, but to obtain education into the latest techniques, moving into one's field of endeavor.

The training steps taken over time should match the needs of the position and also prepare one for the next step in the process of growing one's career. For example, the training that control-room operators of nuclear power plants (NPPs) get seems to be both progressive and meets the requirements of the regulator. However, the depth of training prior to the Three Mile Island (TMI) #2 accident was nowhere as deep as it is now. The analysis of the TMI #2 accident followed what was superficially indicated—that the whole problem was caused by the unpreparedness of the operators, whereas a deeper perception of the circumstances led to realizing the real problem lay with the management and the regulators of the industry. It was their unpreparedness that led to that of the operators. The operators were the front edge of the response to the accident. However, operators do not design systems, see the need for procedures to help the operators, or control budgets. Operators do what they are asked to do! The top managers are responsible for the survival and profitability of the company. Most managers have some technical training and have taken a course in business operation. Although a master's in business administration is useful, it does not fulfill the need to ensure that a company will not go broke because the managers did not run the company to avoid accidents that can cause the demise for the company.

It is suggested that a company should instigate a training process that equips the up-and-coming staff to be able to step into the top manager's job when needed. Although knowing where the money is being spent, how many hours does it take to perform an action, and what skills are needed by the work-staff is important. This kind of work is relatively easy compared to really ensuring that the company continues to exist after being challenged by an accident!

If one learns anything from Adm. Rickover, one can appreciate that putting a submarine out to sea in an unprepared condition is foolhardy. The source of the problem may be equipment failure, failure of personnel from the captain downward, and the uncontrolled leaks of radiation. Adm. Rickover took his responsibility to heart and would take action to prevent such occurrences (see Oliver, 2014).

The top managers and those who will replace them need two things: technical knowledge of the processes and equipment in their organization and responsibility to take actions as needed. Adm. Rickover tried to cover both aspects in his approach to training. Here, we can suggest an approach to the technical side of the equation. As to personnel selection, his approach to this aspect was to select personnel by interview and his tactics were considered by some to be rough. If a candidate did not measure up to Adm. Rickover's requirements, he was out of the Navy submarine world.

In the non-Navy–submarine world, the risk of making the wrong selection may not be not quite as high as a loss of a sub at sea and the corresponding hole in the strategical defense map and the possible loss of a war, but it does represent some degree of elevated risk depending on the industry.

A study of the set of accidents, as described in Chapter 7, shows that there can be significant losses resulting from the poor decisions related to the selection of managers. Managers whose approach is not to pay close attention to the details of the operation can lead to significant losses of both personnel and material.

## 12.3   TECHNICAL TOOLS FOR MANAGERS

In this section, the types of programs needed to cover the technical tools that managers need are listed. Understanding the technical details does not of itself lead to being a good leader and manager. It is suggested that the persons making the selection of a manager should pay close attention to books on the record of the Navy submarine service and Adm. Rickover's approach to the selection of personnel. In addition, it is recommended that close attention be paid to the accidents recorded in Chapter 7 and the roles played by upper management in the cause and accident progression.

The technical tools are given in the following:

1. Beer's viable systems model: This provides insights into the dynamic functioning of organizations.
2. Rasmussen's skill-/rule-/knowledge-based behavior: Gives one an insight into how people respond to different tasks and that one should consider behavior in designing tasks.
3. Ashby's law: Gives one a view of relationship between systems and control requirements.
4. Review of accidents: Provides an insight into how decision-making can affect accident initiation and progression and the feasability of persons selected to be managers.
5. Study of PSA methods: Knowledge can provide some guidance on how accidents can grow and their consequence.

## 12.4   CONCLUSIONS

Management is not all about technology, but it is necessary for managers to be technically competent and quick in understanding problems. Adm. Rickover, who seemed very adept at sizing up persons, placed a great importance on this, plus being responsible and concerned with details. The training process for all high-risk organizations (HROs) must be concerned with ensuring that persons selected for management are concerned about the safety of the public and personnel, and not taking steps that can lead to the loss of the company.

As one can see by reviewing the accidents/incidents, some senior management does not seem concerned with the details; the result is loss of equipment and often personnel. Clearly, those people were unsuited to the positions they held. So, one of the requirements of a company training program is to evaluate the quality of the persons—those who pay close attendance to details—and select the correct person who could make the difference between success and failure of the organization.

# 13 Investment in Safety

## 13.1 INTRODUCTION

The objective of management in any business organization is to optimize operational costs in order for the price of products to remain competitive and thus maintain acceptable profit margins. Shareholders expect the managers of companies to ensure that their focus is firmly on shareholder value by maximizing profit levels. For state-owner organizations, governments expect either a return on their investments to bolster the public purse or at least remain neutrally financed in order not to require subsidies from the public purse. In times of economic downturn or as products begin to approach the end of their "shelf life," profit margins become squeezed and management will inevitably resort to cost cutting. As we have seen with Occidental Petroleum with its Piper Alpha gas production rig and with Bhopal in India, management did not ring-fence safety systems when implementing cost cutting—with tragic results. In their defense, management has argued that there is no method for evaluating how much should be invested in safety and often the "belt and braces" approach leads to vastly exaggerated levels of investment that results in overpriced products that lose market share. They continue: often safety case derivation is produced in isolation of business needs and based on utopian ideologies.

In this chapter, we review the ethos of shareholder value in order to try to understand how a rigorous application can lead to disasters or near disasters. We shall also see how shareholder value has been extended to stakeholder value where stakeholders would include the public, the environment, and worker safety. The second part of the chapter will look at a recent method developed for the UK nuclear authorities, which aims to calculate the required level of investment in safety to meet a prescribed safety case based on the value of human life and the damage that could be caused to the environment.

## 13.2 MANAGING FOR SHAREHOLDER VALUE

In New York at the Pierre Hotel on August 12, 1981, Jack Welch, the newly appointed CEO of General Electric, presented an important new business concept: "Growing fast in a slow-growth economy." This was universally acknowledged by the business world as the "dawn" of the obsession with shareholder value. However, the concept of managing for shareholder value (MSV) had been around for some 30 years to attract institutional investors managing portfolios of shareholdings in large corporations. Briefly, MSV is the value delivered to shareholders as a result of management's ability to grow earnings, dividends, and share price value; it is the sum of all strategic decisions that affect the firm's ability to efficiently increase the amount of free cash flow over time. Making wise investments and generating a good return on capital employed are two important pillars of shareholder value. However, there is a fine line between growing shareholder value responsibly and doing whatever is needed to generate a profit. Aggressively chasing profit at the expense of safety and the environment can cause the concept of shareholder value to fall into disrepute. To blame the MSV concept for all negative outcomes is a mistake as there is nothing in the theory or practice of MSV that forces companies to optimize returns in order to reward owners at the expense of undermining safety and alienating customers, employees, or other stakeholders.

Here we will provide an overview to the workings of shareholder value followed by the linking of overly aggressive application of MSV as a possible contributor the accidents described earlier.

## 13.3   BRIEF OVERVIEW OF THE PRINCIPLES OF MSV

MSV enables managers to develop a granular view of where strategies, activities, and resources add value or subtract from it. Common metrics such as cash flow, return on investment (ROI), economic profit, and total shareholder return allow managers to compare performance across different businesses or product lines, identify and address wasteful or uncompetitive practices, quantify potential growth opportunities and trade-offs, and measure performance outcomes against expectations and against peers. These metrics also ensure that company boards become disciplined about how they allocate capital and evaluate potential investments carefully against the alternative of returning cash to investors. A simple illustration of MSV in action is shown in Figure 13.1.

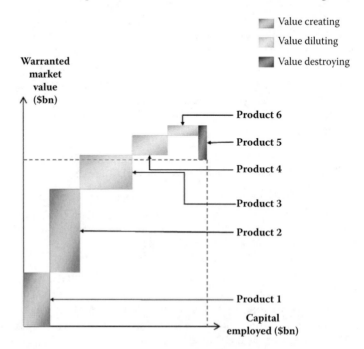

**FIGURE 13.1**   Graphical representation of MSV for a company preadjustment. (Courtesy of PA Consulting Group.)

The efficiency of a company's business can be described graphically as shown with axes representing capital employed and warranted market value. The two axes have normal business meanings; capital employed is usually taken to mean total assets minus current liabilities, and warranted market value refers to the price that prospective buyers might be willing to pay to purchase the company; that is, the worth of the company in the market. Figure 13.1 shows six products or business profit centers belonging to a particular company. Each product (or business profit center) shows the required capital employed against how much each product delivers toward the warranted market value of the company. Products are categorized as value creating, value diluting, and value destroying as shown. It can be seen that Product 5 is value destroying, with another three products being classified as value diluting. Only two products can be considered value creating.

Using the concept of MSV for each product, management exercises its mandate by attempting to elevate the share price by increasing the profit in harmony with a reduction in the capital employed. If undertaken effectively then it may be possible to enhance all products but especially to move value diluting products and value destroying products to a satisfactory value creating profile as shown in Figure 13.2.

The preferred way to reduce capital employed is to reduce the total assets through a lower level of investment in fixed assets such as equipment, plant, or buildings. The most straightforward way to increase profit is to reduce the cost of sales. Reducing this cost has the added advantage of increasing competitiveness thus together the product portion of the warranted market value will increase.

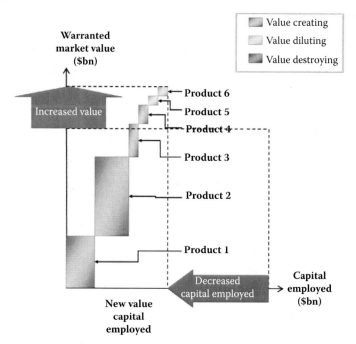

**FIGURE 13.2** Graphical representation of MSV for a company postadjustment. (Courtesy of PA Consulting Group.)

Applied ethically, MSV is an effective way of managing products or business profit centers. If a product does not respond positively to MSV measures management may decide to phase the product out, invest in upgrading the product, or accept for business reasons it remains in a diluting or destroying position. However, as Jack Welch argued, MSV is a strategic management tool and may not be effective if used for short-term management action.

There have been several examples of MSV being used for short-term gains accompanied by unethical management intensions. For a safety-critical business a significant proportion of both the capital employed and cost of sales may be tied to meeting and maintaining the safety case. Reducing the manpower and fixed assets associated with management of the safety case can very quickly convert a diluting or destroying product/business unit into value creating as the sums involved in the safety case could be significant. This action is not only unethical but dangerous to both the natural environment and life, examples being at Bhopal in India, Piper Alpha in the North Sea, and Flixborough in the United Kingdom. Many of the industrial disasters covered in this text and in other texts can have the root cause traced to an unethical use of MSV at its worst or a very naïve application at best. In some cases management remuneration or bonuses may have been associated with improving a company's warranted market value. We should also note that share option schemes rely on an increase in the share price, which relates directly to improvements in the warranted market value of the company.

How much money should a company spend on safety measures to comply with the agreed safety case? Glib answers may include the extremes "as much as we need to" or "the minimum we can get away with"; either one of the extremes will probably not result in a satisfactory outcome for the business. The first could result in a significant overspend on safety, and the second may result in an accident or disaster. Until recently there has been no satisfactory quantitative method through which a sensible investment level in safety can be arrived at. The next section will introduce a quantitative method that has been developed for the UK nuclear power generation industry.

## 13.4  USING THE *J*-VALUE TO ESTIMATE THE LEVEL OF INVESTMENT NEEDED FOR SAFETY

The *J*-value has its origin in the investment in safety at nuclear power plants to reduce human suffering and death caused by unplanned nuclear incidents. Given that economic resources are inevitably limited, the best overall return in terms of suffering averted or lives extended will occur when the regulators impose consistent standards for safety expenditure across all sectors of industry, as recommended by government (HM Treasury, 2005). It was clear that a comparative study would benefit greatly from a scale against which to measure safety investments in different sectors. Building on the life-quality index ($Q$) by Pandey and Nathwani (2003), researchers at City University of London (Thomas, Stupples, and Alghaffar, 2006) devised an absolute scale based on the judgment, or *J*-value, derived in the following, and used this to assess whether any given health and safety expenditure is reasonable. Clearly such an absolute scale can be used for comparative studies also.

Moreover, this method addresses the weakness exposed in the application of MSV in safety-critical industries where only a short-term focus is used.

Later research into use of the $J$-value extended its use into considering damage to the environment, which for safety-critical industries is important given the damage resulting from the Chernobyl nuclear accident. In this section we will provide an introduction to the development of the $J$-value together with an example application.

### 13.4.1  FORMULATION OF THE $J$-VALUE

Let us assume that an average person chooses to spend a part $\Delta G$ ($/year) of his or her yearly income on a scheme that improves safety. As a result, his or her discounted life expectancy will increase by an amount $\Delta X_d$; the life-quality index will change to a new value, $Q' = Q + \Delta Q$, given by

$$Q' = (X_d + \Delta X_d)(G + \Delta G)^q.$$

This is equivalent to the summation of the annual utilities $(G + \Delta G)^q$ spread over the discounted remaining years of life. This expression may be expanded to

$$Q + \Delta Q = G^q \left(1 + \frac{\Delta G}{G}\right)^q (X_d + \Delta X_d)$$

$$= G^q \left(1 + q\frac{\Delta G}{G} + ...\right)(X_d + \Delta X_d)$$

$$= G^q X_d + qG^{q-1}X_d\Delta G + G^q\Delta X_d + ...$$

Assuming that $\Delta G$ and $\Delta X_d$ are small, we may simplify this expression, giving

$$\frac{\Delta Q}{Q} = q\frac{\Delta G}{G} + \frac{\Delta X_d}{X_d}.$$

To be acceptable, the scheme should not yield a net disbenefit so that the previous expression should be equal or greater than zero, which is equivalent to the following condition on the change of income required to fund the project:

$$-\Delta G \leq \frac{G\Delta X_d}{qX_d}$$

Here the minus sign shows income reduction, meaning a person can spend a share of his or her income for an extension to life. This equation implies that the change in annual utility will be felt for the rest of his or her expected life, so that the change in annual income will be experienced over the same period, or

$$X_d + \Delta X_d \approx \Delta X_d.$$

Therefore, the total cost of the safety measure will be $X_d\Delta G$. The maximum reduction in earnings that an average individual should be willing to pay each year corresponds to the limiting case of equality in condition. Now assume that the benefits of the risk reduction are experienced by a population of size, $N$. In this case, the maximum that the population should be willing to pay each year is given by $a_{pop}$ ($/year), which may be calculated from

$$a_{pop} = -N\Delta G = \frac{NG\Delta X_d}{qX_d}.$$

Assume that the actual amount being spent on a safety measures is $\hat{a}_{pop}$ ($/year). Since $a_{pop}$ represents the theoretical maximum, the ratio, $J$ must satisfy the condition

$$J = \frac{\hat{a}_{pop}}{a_{pop}} \leq 1.$$

This equation may be rearranged into criteria for an annual safety spending of

$$J = \frac{qX_d \hat{a}_{pop}}{NG\Delta X_d} \leq 1$$

where $J$ emerges as a judgment variable derived from the life-quality index $q$. The condition $J = 1$ will represent the maximally risk-averse case where the maximum reasonable sum is being spent on safety measures. The $J$-value is the ratio of the actual spending to the maximum reasonable spent, implying that a safety scheme with a $J$-value greater than unity will cause a net disbenefit. Hence, if a scheme is calculated to have a $J$-value of 3.0, then it is costing three times what it should, and effort should be focused on finding another way of producing similar safety benefits more cheaply hence supporting an MSV application. Meanwhile, a safety scheme with a fractional $J$-value is acceptable. For example, if the $J$-value is 0.2, the scheme will yield a good safety benefit without using up too much resource. In fact, the scheme could be augmented and made up to five times as expensive without creating a net disbenefit (although the low $J$-value should not be taken to imply that spending more money is essential).

A severe accident at an industrial plant has the potential to cause, in addition to human harm, general physical and environmental damage and hence expense associated with ground contamination, evacuation of people, and business disruption. Such environmental costs may be comparable with or larger than the cost of direct health consequences, as assessed objectively by the $J$-value approach. While the low probability of the accident may mean that the expectation of monetary loss is small, a published paper (Thomas and Jones, "Extending the $J$-value framework for safety analysis to include the environmental costs of a large accident," 2010) develops a utility-based approach to determine how much should be spent on protection systems to protect against both environmental costs and human harm.

The behavior of the fair decision-maker in an organization facing possible environmental costs is represented by an Atkinson utility function (Atkinson, 1970), which is dependent on the organization's assets and on the elasticity of marginal utility or, equivalently, the coefficient of relative risk aversion, "risk aversion" for short.

$$\delta Z_0 \cong C(\lambda_1 - \lambda_2)T$$

This is a simple expression for the expected reduction in environmental costs from rare accidents with individual offsite cost, $C$, achieved by reducing the frequency of occurrence from a low frequency, $\lambda_1$, to an even lower frequency, $\lambda_2$, for the operating period, $T$, when the growth rate $r_{org} = 0$ is assumed. The maximum sensible spending, $\delta Z_R$, on protection against environmental costs after allowing for risk aversion is calculated as

$$\delta Z_R = M_{R(\varepsilon_{max})} \delta Z_0.$$

A second judgment value, $J_2$, is derived from the investment on the protection system after subtracting the amount sanctioned to prevent direct human harm.

$$J_2 = \frac{\delta \hat{Z}}{\delta Z_R}$$

This net environmental expenditure is divided by the most that it is reasonable to spend to avert environmental costs at the highest rational risk aversion. The denominator in this ratio is found by first calculating the maximum sensible spending at a risk aversion of zero, and then multiplying this figure by a risk multiplier to give the maximum, fair amount to avert environmental costs. The risk multiplier incorporates a risk aversion that is as large as it can be without rendering the organization's safety decisions indiscriminate and hence random.

Risk aversion reflects the decision-maker's reluctance to invest in safety systems. This reluctance to invest is the scaled difference in expected utility before and after installing the safety system and has a minimum at some given value of risk aversion known as the "permission point," and it has been argued that decisions to sanction safety systems would be made at this point. As the cost of implementing a safety system increases, this difference in utility will diminish. At some point, the "point of indiscriminate decision," the decision-maker will not be able to discern any benefit from installing the safety system.

This point is used to calculate the maximum reasonable cost of a proposed safety system. The value of the utility difference at which the decision-maker is unable to discern any difference is called the "discrimination limit." By considering the full range of accident probabilities, costs of the safety system, and potential loss of assets, an average risk aversion can be calculated from the model (Thomas, 2013).

An overall, total judgment value, the $J_T$-value, takes into account the reduction in both human harm and environmental cost brought about by the protection system.

$$J_T = \frac{\delta \hat{W}}{\delta Z_R + \delta V_N}$$

where $\delta \hat{W}$ is the cost of the protection system and $\delta Z_R$ is the maximum justifiable expenditure to save environmental costs—taking account of the maximum value of risk aversion before discrimination is lost, equal to the amount after allowing for disproportion or gross disproportion—and is given as $M_{R(\varepsilon_{max})} \delta Z_0$. $\delta V_N/4$ is the maximum fair amount needed to protect a cohort of $N$ people. Expected reduction of environmental costs is given as $\delta Z_0 \cong C(\lambda_1 - \lambda_2)T$, with $C$ being the cost of one accident and $T$ the length of the interval. Applying the expansion to the $J_T$ formula we get

$$J_T = \frac{\delta \hat{W}}{M_{R(\varepsilon_{max})} C(\lambda_1 - \lambda_2)T + \delta V_N}.$$

$\delta V_N$ is the maximum reasonable spend to protect $N$ people at $J = 1$ and is given by

$$\delta V_N = N \frac{G}{q} \frac{1 - e^{-r_d X_d}}{r_d X_d} \quad \text{for } r_d > 0$$

$$= N \frac{G}{q} \delta X_d \quad \text{for } r_d > 0$$

where $G$ is the GDP per person, $r_d$ is rate of time preference or discount rate, and $q = 1 - \varepsilon$.

## 13.4.2 LIMITING RISK MULTIPLIER

The risk multiplier shows the maximum sum that the organization should be prepared to spend on the environmental protection system at a risk aversion of $\varepsilon$ divided by the maximum $B_D(0)$ it would be prepared to spend at $\varepsilon = 0$. $B_D(0)$ is the break-even cost, equal to the expected monetary saving after the protection system has been installed. If the protection system were to be 100% effective, then $p_2 = 1$, and $B_D(0) = (1 - p_1)C = \pi_1 C$ would be the expected monetary saving brought about by a perfect protection system, equal to the expected loss in the absence of a protection system. In normal case $B_D(0) = (\pi_1 - \pi_2)c$, where $c$ represents a total cost of accident $C$ relative to organization's starting assets A so that $c = C/A$. Here $\pi_1 = 1 - p_1$ is the accident probability without the protection system, and $\pi_2 = 1 - p_2$ is the accident probability after the installation of the protection system.

Both the $J$-value and the $J_T$-value methods make use of concave utility functions of the generalized Atkinson form of the following equation:

$$U_\varepsilon(x) = \frac{x^{1-\varepsilon} - a}{1 - a\varepsilon} = \log x$$

for $0 \leq \varepsilon < 1$ when $a = 0$
for $\varepsilon \geq 0$ but $\varepsilon \neq 1$ when $a = 1$
for $\varepsilon = 1$ when $a = 1$
where $a$ is a binary variable, which produces a power utility function when it is 0 but gives the Atkinson utility function (Atkinson, 1970) when it takes the value 1, as used by H.M. Treasury in the evaluation of public schemes (H.M. Treasury, 2009). In the equation above, $x$ is the money, either wealth ($) or income ($ × year−1), $\varepsilon$ is the modulus of the elasticity of marginal utility of wealth or income, $U_\varepsilon(\cdot)$ is the utility operator at a given value of $\varepsilon$, and the log is taken as the natural logarithm.

By the definition, modulus of the elasticity of marginal utility in simple terms can be referred to as risk aversion. Risk aversion is the reluctance of a person to accept a bargain with an uncertain payoff rather than another bargain with a more certain, but possibly lower, expected payoff. For example, a risk-averse investor might choose to put his or her money into a bank account with a low but guaranteed interest rate, rather than into a stock that may have high expected returns, but also involves a chance of losing value. In large organizations, such as commercial companies, most decisions will be taken on the basis that it is only the money that is of concern to them, implying risk neutrality. But given both the sensitivity of large organizations to criticisms of their safety culture and the requirements of the regulator, some organizations may indeed be displaying a relatively high degree of risk aversion with regard to expected environmental costs.

This may be the case for nuclear plant operators in particular, where public qualms are perceived to be high or financial organizations for which reputational loss may lead to irreparable consequences. If there is no protection system and we are not expecting any accidents to occur, the organization's utility is given as $u_0(\varepsilon) = U_\varepsilon[A]$.

The expected utility in the case where there is no environmental protection is then

$$E(u_1) = p_1 U_\varepsilon[A] + (1 - p_1) U_\varepsilon[A - C] = U_\varepsilon[A - C] + p_1 (U_\varepsilon[A] - U_\varepsilon[A - C])$$

where $u_1$ (the utility without the environmental protection system) is a random variable that can take values of either $U_\varepsilon[A]$ or $U_\varepsilon[A - C]$, while $E(\cdot)$ is the expectation operator. If the environmental protection system is installed for cost $B$, then the new utility, $u_2$, will be a random variable that may take either of the values, $U_\varepsilon[A - B]$ or $U_\varepsilon[A - B - C]$, and will have an expected value given by

$$E(u_2) = p_2 U_\varepsilon[A - B] + (1 - p_2) U_\varepsilon[A - B - C]$$
$$= U_\varepsilon[A - B - C] + p_2 (U_\varepsilon[A - B] - U_\varepsilon[A - B - C]).$$

The utility difference $E(u_1) - E(u_2)$ at a given value of risk aversion is then will be given by

$$D(u_1, u_2 \mid \varepsilon) = \frac{A^q}{q_a} \{ [1 - c]^q - [1 - b - c]^q + p_1 (1 - [1 - c]^q) $$
$$- p_2 ([1 - b]^q - [1 - b - c]^q) \}$$

for $\varepsilon \geq 0$ and $\varepsilon \neq 1$
where $b = B/A$, $c = C/A$, $q = 1 - \varepsilon$, $q_a = 1 - a\varepsilon$. Comparing the expected utility difference at a given value of risk aversion with the utility of the starting assets we get

$$R_{120}(\varepsilon) = \frac{D(u_1, u_2 \mid \varepsilon)}{u_0(\varepsilon)}.$$

Value of $R_{120}(\varepsilon)$ will depend on utility difference value since utility of the starting assets will always be positive assuming that assets are greater than zero. A positive value of $R_{120}(\varepsilon)$ indicates the normalized loss expected in the organization's utility. Thus $R_{120}(\varepsilon)$ provides a convenient, dimensionless measure of how reluctant the decision-maker will be to invest in a protection system. Hence we may denote $R_{120}(\varepsilon)$ the "reluctance to invest." A 100% reluctance to invest, equivalent to a point-blank refusal to invest, will be associated with a protection system that is so expensive that it is expected to reduce the utility of the organization's assets to zero at that risk aversion, $\varepsilon$.

The negative, $-R_{120}(\varepsilon)$, will be termed the "desire to invest." For the Atkinson utility function, $a = 1$, and so the reluctance to invest is now $R_{120A}$, given by

$$R_{120A}(\varepsilon, A) = \frac{D(u_1, u_2 \mid \varepsilon)}{u_0(\varepsilon)} = \frac{A^q}{A^q - 1} \{ [1-c]^q - [1-b-c]^q $$
$$+ p_1(1 - [1-c]^q) - p_2([1-b]^q - [1-b-c]^q) \}.$$

In part 2 of their paper (Thomas, Jones, and Boyle, "The limits to risk aversion," 2010) the authors state that there must be a value of $\varepsilon$ at which $R_{120}$ will reach a minimum, when it is at its most negative. This point of maximum negativity in the reluctance, $R_{120}$, will correspond to the point of maximum desire to invest in the protection scheme.

The corresponding value of the risk aversion, $\varepsilon = \varepsilon_{pp}$, will be the one at which the decision-maker will see most clearly the desirability of implementing the environmental protection system and will now grant permission for it to be installed. Hence we may call $\varepsilon_{pp}$ the "permission point." Using the Atkinson utility function, the permission point $\varepsilon_{pp}$ will depend on the assets, $A$; the prior probability of the loss not happening, $p_1$; the posterior probability of the loss not happening, $p_2$; the normalized cost of the loss, should it happen, $c$; and the normalized cost of providing full protection against the loss, $b$.

Figure 13.3 shows the dependence of reluctance to invest on the value of risk aversion (Risk Aversion Calculations, 2014). In fact, we shall find it convenient to use in place of $b$ a normalized version, namely the risk multiplier $M_R = b/B_D(0)$. Therefore, knowing the maximum sensible value of risk aversion we may calculate the risk multiplier. The new $J_T$-value will show similar behavior to the original $J$-value, in that $J_T < 1$ will indicate reasonable value for money, while $J_T > 1$ will indicate overspending on protection that will need to be justified by further argument. While the analysis is phrased in terms of environmental costs, the treatment is sufficiently general for all costs, including onsite damages, loss of capability, etc., to be included. The new $J_T$-value method provides for a full and objective evaluation of the worth of any industrial protection system.

### 13.4.3    APPLICATION OF THE *J*-VALUE

A worked example using the *J*-value was presented in Thomas et al. (2010). A company with assets of $A = \$10$ billion owns and operates a nuclear power plant with 50 years of life remaining. It is considering installing an additional protection system that will reduce the frequency of a large accident from $\lambda_1 = 2 \times 10^{-5}$ year$^{-1}$ to $\lambda_2 = 5 \times 10^{-8}$ year$^{-1}$. The new protection system would last the life of the plant and cost $\delta \hat{W} = \$4.5$ million, a sum that includes all finance and maintenance charges. The company will be aware of MSV and will aim to cap this figure.

A risk analysis carried out by the company has shown that if the large accident were to occur, then five of the company's workers could be expected to be killed outright, while 40 would be exposed to a one-off dose of 300 mSv. Moreover, 500 members

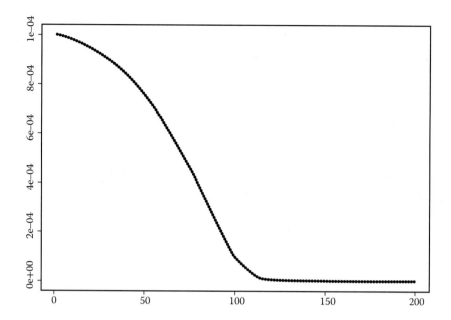

**FIGURE 13.3** Reluctance to invest vs. risk aversion.

of the general public living in a small town close to the plant would receive a one-off dose of 200 mSv, while the remaining 5,000 inhabitants of the same town would receive a single dose of 150 mSv. In addition, there would be environmental costs of $C = \$5$ billion, covering evacuation, relocation, business disruption, decontamination and cleanup, among others. Take the net discount rate $r_d$, as zero, but assume that the growth rate of the organization, $r_{org}$, is 2% per year. The owner of the plant is prepared to accept a disproportion factor of three in his or her evaluation of human harm: $J^* = 3$. Should the protection system be installed?

### 13.4.3.1 *J*-Value Analysis

The average loss of life expectancy across the cohort of 5,545 people affected, workers and public, $\varepsilon_{dl}$, is 0.4 years as calculated from the CLEARE program (change of life expectancy from averting a radiation exposure); the results are summarized in Table 13.1 (based upon actuarial data for the United Kingdom for 2007).

The expected improvement, $\delta X_d$, in discounted life expectancy arising from the installation of the protection system, is given by

$$\delta X_d = \left(\lambda_1 - \lambda_2\right) M \varepsilon_{dl}$$

where $\varepsilon_{dl}$ is the change in discounted life expectancy caused by a single incident, $\lambda_1$ is the accident frequency without the scheme ($2 \times 10^{-5}$ in this case), while $\lambda_2$ is the frequency with the scheme ($5 \times 10^{-8}$), while $M = 50$ years is the duration of protection. With a zero discount rate assumed, the life expectancy and discounted life expectancy are the same, so that $\delta X_d = 3.99 \times 10^{-4}$ year. The maximum reasonable spending to protect $N$ people at $J = 1$, $\delta V_N$, is given by

**TABLE 13.1**

**Loss of Life Expectancy Given That the Accident Has Happened**

| Group | Group Size | Dose (Sv) | Loss of Life Expectancy per Person (y) |
|-------|-----------|-----------|----------------------------------------|
| Public | 5000 | 0.15 | 0.354 |
| Public | 500 | 0.20 | 0.472 |
| Plant operators | 5 | killed outright | 38.795 |
| Plant operators | 40 | 0.3 | 0.401 |
| Average loss of life expectancy per incident, per person, across all groups = | | | 0.400 |

$$\delta V_N = N \frac{G}{q} \frac{1-e^{-r_d X_d}}{r_d X_d} \delta X_d \quad \text{for } r_d > 0$$

$$= N \frac{G}{q} \delta X_d \quad \text{for } r_d = 0$$

where $G$ is the GDP per person, taken as the 2007 figure of $22,997, and $q = 1 - \varepsilon$, in which $\varepsilon = 0.82$, the value appropriate for safety trade-offs. Selecting the lower line in equation directly earlier because $r_d = 0$, and putting $N = 5,545$, gives

$$\delta V_N = £282,386.$$

This would give a $J$-value of $4,500,000/282,386 = 15.9$. Since $J > 1$, the protection scheme cannot be justified based on protection against human harm alone: the accident is severe, but it is already of low frequency. The disproportion factor of three will raise the maximum reasonable expenditure to

$$\delta V_N \big|_{J^*=3} = £846,919$$

but even so, this will justify only about a fifth of the cost of the protection system. However, the protection system will protect against very substantial environmental costs also. Thus

$$\delta Z_0 = \frac{1-e^{-r_{org}T}}{r_{org}T}\left[K_c(\lambda_2)e^{-\lambda_2 T} - K_c(\lambda_1)e^{-\lambda_1 T} - \left(K_c(\lambda_2) - K_c(\lambda_1)\right)\right]$$

gives the justifiable spend to reduce the frequency of environmental damage at the risk-neutral point (a risk aversion of zero):

$$\delta Z_0 = £3,165,746$$

To find the risk multiplier, $m_r(\varepsilon_{max}) \approx M_R(\varepsilon_{max})$, we note that the probability of the accident occurring in 50 years is $\pi_1 = 1 - p_1 \approx \lambda_1 M = 2 \times 10^{-5} \times 50 = 10^{-3}$. We will assume, conservatively in terms of the risk multiplier, that the protection system will eliminate the possibility of the accident, so that $\pi_2 = 1 - p_2 = 0$. The problem is thus the same as that treated in Example 1 of Thomas, Jones, and Boyle (2010b), for which it was found that $m_r\left(\varepsilon_{max}\right) = 1.34$. Hence

$$\delta Z_R = M_R\left(\varepsilon_{max}\right)\delta Z_0;$$

it follows that

$$\delta Z_R = \pounds 4,242,100.$$

Comparing $\delta Z_R$ and $\delta \hat{W}$, it is clear that the protection system is not quite justified on environmental cost avoidance alone, since

$$J_{20} = \frac{\delta \hat{W}}{\delta Z_R} = \frac{4,500,000}{4,242,100} = 1.06.$$

(Even though a practical decision-maker would almost certainly accept this as close enough). However, the $J_2$-value, which allows for human harm, including the disproportion multiplier, of $J* = 3$ is in this case

$$J_2 = \frac{\delta \hat{W} - J* \delta V_N}{\delta Z_R} = \frac{4,500,000 - 846,919}{4,242,100} = 0.86.$$

Meanwhile, the $J_T$-value comes out as

$$J_T = \frac{\delta \hat{W}}{\delta Z_R + \delta V_N} = \frac{4,500,000}{4,242,100 + 282,306} = 0.99.$$

Thus $J_T \leq 1$, and the protection system should be installed.

## 13.5 CONCLUSION

A misuse of MSV can lead some company management to optimize shareholder value by manipulating capital employed for short-term gains. The aim could be to optimize the share price to benefit share-option holders or to improve the market value of the company in preparation for sale or takeover. In safety-critical business this misuse of MSV could lead to unjustified cost reductions in elements of the business that relate to the safety case. More able managers would view safety as a critical activity within the company and would seek to identify an investment level in safety that is allied to the risk exposure of the business. Until recently there has been no effective way to judge the level of asset investment required to meet the safety case and this has led to both over- and underinvestment, but more importantly opened up safety to cost-cutting actions with some tragic consequences. A judicious application of a $J$-value calculation could be used by management to set an adequate level for the investment in safety, and by regulators to ensure that the investment is maintained and not maladjusted to manipulate MSV.

# 14  Conclusions and Comments

## 14.1  INTRODUCTION

In this book, we have introduced a number of concepts to help improve management decision-making for high-risk organizations (HROs) under stress. Our objective in putting together a set of concepts is to improve the management processes. The concepts that we have introduced are a cybernetic organization model based upon the functions of the human body, namely cognitive processing; manual actions to be taken; the automatic action (controls); feedback mechanisms; learning from past experiences; input information derived from sensors, etc. We introduced Beer's model, viable systems model (VSM), to mimic the characteristics of an industrial organization. The organizational model is then linked to the utility/plant model, the external and internal disturbances, and the regulators, so one can understand how all these things function together: what disturbances can lead to accidents and the responses of the organization and the plant are, given the accident initiators.

We have reviewed the concepts of Adm. Rickover relative to running the US nuclear Navy and added to them, based upon technology and ideas that have occurred after his time. The book has explanations of these methods and techniques in earlier chapters. Adm. Rickover's approach is a framework or foundation upon which one can build, and his seven principles can be seen in the Appendix.

Effectively, the organization needs to work together to run a plant effectively (economically) and safely. The organizational functions corresponding to the cybernetic model are the CEO as the brain, and the lower-level managers and the operators, for example, like the hands taking actions at the direction of the brain. One can then appreciate that all members of the organization needs training to fulfill their functions.

We have covered the technical aspects of trying to improve aspects of managerial control of operations by concentrating on education and training, but we are not insensitive to economics. We realize that for an organization to be viable, managers must pay attention to cost control. Chapter 13 does cover these considerations, so it is important to balance the technical side with the cost of carrying out technical improvements. This balance is important to the survival of the organization. Our review of the Northeast Utilities operation showed that one cannot just cut manpower and end up with an effective and safety-conscious organization. Improvements can be made in the organization to reduce costs and still have an effective organization, but it does call for management to pay close attention to use of technology to enable this to be done.

## 14.2   ANALYTICAL ELEMENTS

The factors that can improve management decision-making are listed in the following and management/staff should have an understanding of these:

1. **The dynamics of organizations.** We selected Beer's model based on the human body. Without understanding how an organization operates and its role and the limitations of the individual parts, management cannot direct the organization to control situations like accidents.
2. **Ashby's law of requisite variety.** How to acquire an understanding of what sources need to be addressed to acquire this information and how the requisite variety can change during accidents. One needs to know what the variety is of the system to be able to control it!
3. **Risk of operations.** This is gained by examining probabilistic risk assessments (PRAs). Risk does convert to what might happen, because it is a probabilistic relation. But it gives one an idea of how often it might occur and the corresponding consequence. As the Fukushima tsunami indicated, even a 1 in 1,000 event may occur the next day!
4. **Human behavior characteristics.** Personnel need to receive the correct training to deal with certain tasks. They will gain an appreciation of what it takes to address a variety of tasks, that is, procedural support, time to solve problems, and knowledge and background.
5. **What it takes to respond to accidents and situations.** This means that managers need a deep knowledge about the various systems with which they are involved or access to advice of knowledgeable persons to compensate for their own lack of relevant information.
6. **Analysis of accidents in the nuclear and other industries.** This is to gain insights into the decisions of other organizations and how to avoid these errors.
7. **Regulators' approach and analysis of plant-related incidences.** How to benefit from these analyses to improve the organization's operations.
8. **Power of simulation techniques.** For training, accident analysis, and understanding the underlying variety and changes that can occur with accidents.

Each of the previous items has been discussed in depth in Chapters 3 through 11. We have tried to cover many of the needs to satisfy the requirement to prepare management for its roles in decision-making. Without these tools and their application, decision-making can be a random process. The analysis of a number of accidents given in Chapter 7 can be used by managers to understand how accidents occur and the corresponding errors of judgment that can contribute to its progression.

## 14.3   CONCLUSION

The methods introduced and applied by Adm. Rickover achieved a breakthrough in the US Navy submarine service and enabled the risk of operation of nuclear submarines to be small, especially when one compares the US with the Russian submarine service (Oliver, 2014, see p. 3). In the Appendix we list seven of Adm. Rickover's principles. The emphasis of his principles is on training, concepts of responsibility and technical self-sufficiency, etc.

We are proposing augmentation of his principles with our recommendations as outlined previously. In a tight organization like a submarine, it is easy to appreciate the roles of persons. To compensate for the difficulties in appreciating the roles of persons in large organizations we have introduced Beer's dynamic virtual systems model, so it is clear to the whole organization of the role and functions of individuals.

Adm. Rickover's technical sufficiency is supported here by an understanding of Ashby's law and an understanding of the power of simulation methods and a deep knowledge of systems.

One thing that Adm. Rickover did not seem to address is the analysis of the possibility of an accident and why these things are possible and their consequences, although it was known at the time that the Russians had submarine accidents. It is thought that he was aware of the direct effect of a loss of submarine, loss of persons, and the environmental effect of radioactive releases.

The whole issue of safety of organizations and the impact on the society has become more visible, especially when dealing with HROs. The impact of organizational failure to control accidents now influences nations, not just a few local persons. This is not to say that local persons are unimportant, but rather the effects of some accidents can affect countries distant from the site of the accident. The Chernobyl accident was observed by persons in Sweden and the effects of radiation were felt as far away as Wales, leading to the slaughter of sheep and lambs because the grass that they ate was contaminated! It is our belief that there should be an understanding not only that an accident has taken place, but the effect of these accidents on people and organizations should be understood from the environmental and the economic cost point of view. Chapter 13 pays attention to cost control that has to be taken into account by organizations. That is, what is the cost of improvements, are they economic to carry out, and do they achieve the safety objectives set out by management and confirmed by regulators?

A central part of the book is the space given to the analysis of accidents in different fields—nuclear, chemical, space, oil and gas, and railways. The motive here is to indicate the role of decision-makers in the whole business of accidents, from initiation to progression and a failure to terminate the progression. As the Beer model showed, the brain of the organization is the central cognitive process and therefore is at the center of the decision process. An operator may step in and do something brave and effective, but it is not his or her job to pick the right steam generator (see Section 7.9.3); it is not his or her job to insist the blowout preventer (BOP) oil-cutoff valve must work (see Section 7.6.1); and it is not his or her job to launch the *Challenger* (Section 7.8.1).

It was the aim of the authors to cover a number of accidents in different fields is to show that accidents are not limited to a particular industry or type of organization, but they are more connected to another common aspect that is not normally closely considered—the chief executive officer (CEO) along with his or her staff and the board of directors, who represent the investors. Remember Adm. Rickover's principle #6, "Require adherence to the concept of total responsibility!" Adm. Rickover also made the following comment at a presentation at Columbia University: "The man in charge must concern himself with details" (Oliver, 2014). Responsibility is not something that can be given away.

The other objective of the set of accidents described in this book was to provide a library of accidents that people can study as part of a preparation for the position of a manager in a high-risk industry and remember what might be the end-state of one of your decisions if taken lightly. The accident in Hungary was one such lightly taken decision (see Section 7.9.2). The Paks NPP organization was a forward-thinking organization and had devoted a lot of attention to the field of risk assessment (PSA) and this accident was out of character.

1. Require rising standards of adequacy.
2. Be technically self-sufficient.
3. Face facts.
4. Respect even small amounts of radiation.
5. Require relentless training.
6. Require adherence to the concept of total responsibility.
7. Develop the capacity to learn from experience.

It is our and others' opinion that Adm. Rickover had a clear objective of what he was trying to achieve, which was to form an advanced submarine force based upon nuclear power plants that were reliable, safe, and at the same time were in advance of any Navy either friendly or unfriendly. He accomplished his objective; some may complain about how he did it, but trying to take a conservative organization and change its direction is no mean feat.

The concept of a nuclear submarine that was a real submarine, not a modified surface vessel, was revolutionary, and when tied to sea-launched ballistic missiles was a complete change in warfare. The associated hunter submarine driven by a nuclear power plant was also revolutionary.

Apart from the nuclear plant, the other part of the story was the attention paid by Adm. Rickover to safety of the submarine and of the crew. Adm. Rickover enforced the idea that all of the equipment was to be highly reliable, and radiation exposure of the crew was to be guarded against. One can see the necessity of this, if the submarine is going to sea for months and the crew is confined in the body of the submarine. Clearly, one cannot stop to change a piece of equipment at sea or be concerned over the radiological leakages during the period at sea. If the health of the crew is challenged, their performance over time is likely to be reduced, which is not what Adm. Rickover or the Navy wanted.

# Appendix: Admiral Rickover's Management Principles

## A.1  INTRODUCTION

This appendix refers to the work of Admiral Hyman G. Rickover. The reason why the work of Adm. Rickover is discussed is because of his influence on the US nuclear industry, both by virtue of the personnel he (or his organization) trained that entered the nuclear utility business, and because of the influence of his philosophy on the leadership of the Institute of Nuclear Power Operations (INPO), who had previously served as admirals in the nuclear Navy.

His contribution to the building of the Navy nuclear program is well known, but he also evolved a management process very specifically addressing the needs of a program that focused on safety issues related to operating nuclear power plants. In his case, the nuclear power plants were those needed to run submarines. Before the invention of nuclear power plants, submarines were really surface ships that occasionally submerged. After the use of nuclear power plants, submarines became real submarines that occasionally came to the surface. Their time under the sea was limited by provisions and the needs of the crews.

However, before submarines could operate this way, the safety of the power plants had to be assured and the crews protected from the possible effects of radioactivity. Adm. Rickover was very concerned with both of these items. He tackled these two issues in several ways, by attention to having reliable equipment and by the selection and training of Navy personnel. Many things that he did caused him not to be liked, but for a large time his success enabled him to proceed with his approach; for instance, he gained the support of the US Senate, which insisted that he be retained when the Navy wished to retire him.

One of the issues that is really important is the safe operation of nuclear reactors. His first insistence was that submarines not be sent out on duty until they were considered to be safe to operate. This often led to conflicts between him and the operational personnel of the Navy. The operational part of the Navy had its duty areas, places to set "boomers," and patrol zones for attack submarines. The conflicts mirror the dichotomy between economics and safety in the case of nuclear power plant (NPP) operations.

The management methods of Adm. Rickover have led to the safe operation of the Navy's submarines and are worth considering, seeing if it is possible to draw on them to shape the civilian NPP programs. In some way, Adm. Rickover's training of Navy personnel has already had an influence on the utility NPP business in that many Navy personnel have joined the commercial NPP business and also INPO has been staffed by Adm. Rickover-trained persons, including a number of admirals.

**159**

## A.2   ADM. RICKOVER'S PRINCIPLES

In the review of "General Public Utilities Nuclear Corp organization and senior management and its competence (after the Three Mile Island 2nd unit accident in March 1979) to operate TMI-1" by Adm. Rickover (see Rickover, 1983), he states his principles of operation in the form of management objectives:

> Require rising standards of adequacy.
> Be technically self-sufficient.
> Face facts.
> Respect even small amounts of radiation.
> Require relentless training.
> Require adherence to the concept of total responsibility.
> Develop the capacity to learn from experience.

In his words, "these principles express attitudes and beliefs. They acknowledge complex technology and that safe nuclear operation requires painstaking care." He also points out senior management must be technically informed and be personally familiar with conditions at the operating plants.

Of course, others may have different ideas and priorities for each of his objectives. Some things like "relentless training, total responsibility, and learning from experience" do stick out. Radiation control and understanding of the consequence of radiation exposure and release should be part of everyone's training, if involved in NPP operation. INPO has adopted much of Adm. Rickover's philosophy, but this is not too surprising, since many of the top personnel come from the nuclear Navy.

## A.3   CONSIDERATION OF ADM. RICKOVER'S PRINCIPLES

One must recognize the value of Adm. Rickover's contributions. It is clear in an organization that not all of the personnel are equal in terms of their impact on the viability of the organization. The leaders in an organization provide the guidance and decision-making for the organization. They provide the direction for the whole, but there are others who provide guidance for others and information to the leaders on the whole of how the operation is functioning. The leader cannot do everything and needs support at all levels within the organization.

Some of the key roles that Adm. Rickover played were those of selecting the officers and also ensuring the scope of training that the crews should undertake. He also maintained pressure on shipyards and suppliers by insisting on quality products. Adm. Rickover believed in progressive training and with increasing responsibilities. The result of this process was to produce technically competent personnel capable of making good decisions and being responsible for their decisions. One advantage that Adm. Rickover had over, say, leaders in the utilities was the tightness of the Navy organization. The personnel signed up as volunteers and wanted a career in the Navy, particularly the nuclear Navy. This is not to say that utility personnel are not dedicated, but that the organizations are much looser and opportunities exist elsewhere at different utilities/organizations. So the culture of the nuclear Navy was, and is, different from that of civilian organizations.

In the matter of training, Adm. Rickover had a number of land sites where operating reactors were used for training purposes. In this way, Adm. Rickover gained a duplicate power module to that existing within a submarine. How more realistic can one get? At that time, he was not in favor of mathematical simulators, which he believed were incapable of preparing his crew to the stresses induced when things went wrong. However, during his review of TMI Unit #1, he must have realized that it was much more difficult to have duplicate NPPs for training purposes and also the quality of NPP simulators had improved to make them more acceptable for personnel training. The simulators are more realistic today than in the early days of nuclear power.

## A.4  CONCLUSIONS

Looking at nuclear utility operations, how should one interpret Adm. Rickover's principles? The first thing that comes to mind is Adm. Rickover's emphasis on safety and quality of equipment used. Clearly, these should be the same for NPP operations as for submarines. Failure of either personnel or equipment can lead to a "lost" mission, failing to supply power to the public and impacting the environment, along with the potential of large economic loss for the utility itself. The unsafe operation of a nuclear power plant in a submarine can lead to the loss of the crew; an environmental disaster depending on where the accident occurs; and potentially the loss of a war, in the case of combat.

The unsafe operation of an NPP can lead to some deaths of personnel, loss of power generation capability, a huge economic cost for the utility, and a significant cleanup problem for society/utility along with possible radioactivity release, depending on the subsequent actions of the utility and government. One thing that is an advantage to prevent the spread of radioactive materials is the containment and its support systems!

In the case of the NPP utilities, their operations are monitored by the NRC and by INPO. For these two organizations, the processes and actions are quite different and these processes are discussed elsewhere in this book. The NRC's actions are regulatory and reactive. Those of INPO are proactive and can be somewhat limited depending on the response of the utility management. Responsibility for safe operations lies principally with the utility.

In the case of the nuclear Navy, the captain of a submarine is the responsible officer at all times for the safe operation of the submarine. If anything happens, then he is empowered to act and if he fails so to do, then it is likely that he would have to leave the service. The responsibility of the chief nuclear officer (CNO) in the case of a utility is less well designated, so what is expected in the nuclear Navy does not necessarily carry over to a utility. The NRC may find the utility at fault in an accident and not hold the CNO, president, or CEO individually responsible, unless it can be proved that it was criminally inspired.

The case studies found that often decisions by CNOs, presidents, and CEOs can provide the environment under which accidents can occur. One thing that is important is the education and training of personnel to perform tasks from simple to complex. The decision to reduce training and education to save money can lead to accidents or near accidents. There is a cost of doing business. Like the captain in the Navy, the CEO and CNOs have the responsibility to ensure the quality of the training programs. This also holds for the quality of materials and the maintenance operations. Also, the selection of personnel is important, since the utility needs good-quality and trained personnel to perform exacting tasks. In fact, this might be more important in the diffused atmosphere of a NPP than in the close confines of a submarine.

One thing that has seemed to come to the surface over the last few years is the safety culture of an organization. It is supposed that the nuclear Navy has a uniform safety culture by view of the training program and the fact that its personnel are volunteers in the military. Schein in his lectures to INPO (Schein, 2003) refers to subcultures as being important, because of problems arising from different views of one's job, how one is paid, and what is expected. One would think that in the nuclear Navy, there could be some cultural problems between executives (captains and above), officers, and ratings. Interestingly, Adm. Rickover did not make any comments on this topic and it does not turn up to be an issue in his review of General Public Utilities Nuclear (GPUN) management of TMI Unit #1!

Although Adm. Rickover contributed much in terms of developing the US nuclear Navy and the associated management organizational, which focused on safety and radiation control, he was not without his critics. Among the charges, were that his tight control led to a failure to develop other submarines, such as those like the Russian Alfa submarines that were faster and dived deeper than US designs. He was also charged with eliminating serious competitors (see comments by Schratz [1983]) in a review of the book *Rickover* by Polmar and Allen, published by Simon and Schuster (1982). Adm. Rickover served for 63 years and that was too long in the eyes of many, especially in the key position that he held!

One of the prime issues associated with management is how one removes top managers when they are found not to advance the safety or competiveness of an organization. Eventually Adm. Rickover was removed or retired from active duty, but that was hard to accomplish. Utility organizations rely on the board of directors to ensure that presidents and CEOs do function correctly relative to the health of the utility from a safety or economic standpoint. Sometimes, the boards of directors fail to perform as required (see MacAvoy and Rosenthal, 2005) to protect the shareholders, employees, and the public from the poor decisions of management.

# References

Aberfan Disaster Tribunal. 1966. Available at: Nuffield.ox.ac.uk/politics/Aberfan/tri.htm. Tribunal started 10/1966.

Aceh Tsunami and Indian Ocean Earthquake. 2004. Available at: Wikipedia.org/wiki /2004-Indian_Ocean_Earthquake_tsunami.

Al-Ghamdi, S. H. 2010. *Human Performance in Air Traffic Control Systems and Its Impact on Safety*. PhD dissertation, City University, London, UK.

Ashby, W. R. 1956. *An Introduction to Cybernetics*. London, UK: Chapman and Hall, Ltd. (First published and reprinted as a paperback, a number of times.)

Atkinson, A. B. 1970. On the measurement of inequality. *Journal of Economic Theory*, 2(3), 244–263.

Beer, S. 1979. *Heart of the Enterprise*. Chichester, UK: John Wiley and Sons.

Beer, S. 1981. *Brain of the Firm*. Chichester, UK: John Wiley and Sons.

Beer, S. 1985. *Diagnosing the System for Organizations*. 6th edition. Chichester, UK: John Wiley and Sons.

BOP. 2015. Energy Photos. Available at: http://www.energyindustryphotos.com.

BP. 2010. *Deepwater Horizon, Accident Investigation*, Report. London: BP.

Braun, M. 2011. "The Fukushima Daiichi Incident," Fukushima Engineering Presentation, AREVA NP, GmbH available at http://hps.org/documents/areva_japan_accident_ 20110324.pdf.

Broadribb, M. P. 2006. BP Amoco Texas City incident. *American Institute of Chemical Engineers, Loss Prevention Symposium/Annual/CCPS Conference*, Orlando, FL.

Carnino, A., Nicolet, J.–L., and Wanner, J.–C. 1990. *Man and Risks Technology and Human Risk Prevention*. New York, NY and Basel, Switzerland: Marcel Dekker.

CNN. 2011. "Expert: Japan Nuclear Plant Owner Warned of Tsunami Threat," CNN Wire Staff, March 28, 2011.

Commander of the Marine Corps. 2011 *United States Marine Training Manual*. Unit Training Management Program, order 1553 3B, Washington, DC: Department of the Navy.

Cullen, W. D. 1990. *The Public Inquiry into the Piper Alpha Disaster*. Vols. 1 and 2. London, UK: HMSO. (Presented to the Secretary of State for Energy 19-10-1990 and reprinted in 1991 for general distribution.)

CSB. 2014. *Explosion and Fire at Macondo Well*. U.S. Chemical Safety and Hazard Investigation Board. Report 2010-10-1-05, Vols. 1 and 2. *New York Times*, 2015.

Danilova, N. 2014. Integration of Search Theories and Evidential Analysis to Web-wide Discovery of Information for Decision Support. PhD thesis, City University, London, UK.

Dudorov, D., Stupples, D., and Newby, M. 2013. Probability analysis of cyber attack paths against business and commercial enterprise systems. *2013 European Intelligence and Security Informatics Conference*, Uppsala, Sweden.

Espejo, R, and Harden, R. (eds). 1989. *The Viable System Model: Interpretations and Applications of Stafford Beer's VSM*, Chichester, UK: Wiley.

Espejo, R. 1993. Strategy, structure, and information management. *Journal of Information Systems*, 3(1), 17–31.

Eyeions, D. A., Seyfferth, L., and Spurgin, A. J. 1961. Analogue computer studies of heat exchangers. *Analog Society Conference*, Opatija, Slovenia.

Flixborough. 1974. Flixborough Disaster. Available at: wikipedia.org/wiki/Flixborough _disaster.

Frank, M. V. 2008. *Choosing Safety: A Guide to Using Probabilistic Risk Assessment and Decision Analysis in Complex, High Consequence Systems.* Washington, DC: RFP Press.

Govan, F. 2013. Dozens killed as train derails in Northern Spain. *The Telegraph.* London, UK.

HAEA. 2003. *Report to the chairman of the Hungarian Atomic Energy Commission on the authority's investigation of the incident at Paks nuclear power plant on 10 April 2003.* HAEA Budapest, Hungary.

Halifax Explosion. 1917. Available at: www.cbc/halifaxexplosion and Wikipedia/Halifax Explosion.

Hannaman, G. W. and Spurgin, A. J. 1984. *Systematic Human Action Reliability Procedure (SHARP).* EPRI NP-3583. Palo Alto, CA: Electric Power Research Institute.

Herring, C. and Kaplan, S. 2001. *The Viable System Model for Software.* Report. Brisbane, Australia: Department of Computer Science and Electrical Engineering, University of Queensland.

HM Treasury. 2005. *Managing Risks to the Public: Appraisal Guidance.* London: HM Treasury.

HM Treasury. 2009. *Managing Risks to the Public: Appraisal Guidance.* London: HM Treasury.

INPO. 2011. *Special Report on the Nuclear Accident at the Fukushima Daiichi Nuclear Station,* INPO 11-005, Rev 0. Atlanta, GA: Institute of Nuclear Operations.

Japan Fire Department. 2011. *FDMA Situation Report, no 135.* Available at http://www.earthquake-report.com.

Joksimovich, V. and Spurgin, A. J. 2014. Issues associated with the closure of San Onofre NPP. San Diego Institute of Electric and Electronic Engineers meeting, February 26, 2014, San Diego, CA.

Kemeny, J. G. 1979. *The Report to the President on the Three Mile Accident.* Originally published October 30, 1979.

Leveson, N. 2004. A new accident model for engineering safer systems. *Safety Systems,* 42(4), 237–270.

Leveson, N. 2011a. *Engineering a Safer World, Systems Thinking Applied to Safety.* Cambridge, MA and London, UK: The MIT Press.

Leveson, N. 2011b. The use of safety cases in certification and regulation. *Journal of System Safety,* 47(60), 13–23.

Link. 1942. The 1942 model C-3 Link Trainer. Western Museum of Flight. Available at: www.wmof.com.

Lydell, B. O. Y., Spurgin, A. J., and Moieni, P. 1986. *Human Reliability Aanalysis of Backflush Operations at Barsebeck, NPP.* NUS Report 4911. For the Swedish Regulator (SRK).

MacAvoy, P. W. and Rosenthal, J. 2005. *Corporate Profit and Nuclear Safety: Strategy in 1990s.* Princeton, NJ: Princeton University Press.

Maeda, R. 2011. Japanese nuclear plant survived tsunami, offers clues on safety. *Reuters,* Oct 21, 2011.

McCurry, J. 2015. Fukushima operator 'knew of need to protect against tsunami but did not act'. *The Guardian.* 18th June, 2009.

NRC. 2016. *Backgrounder: Probability Risk Assessment.* Washington, DC: NRC Office of Public Affairs Operation.

NRDC. 2011. *The BP Oil Disaster at One: A Straightforward Assessment of What We Know, What We Don't Know and What Questions Need to Be Addressed,* National Resources Defense Council, NRDC Report. Available at http://www.nrdc.org/energy/bpoildisarsteroneyear.asp.

Oliver, D. 2014. *Against the Tide, Rickover's Leadership Principles and the Rise of the Nuclear Navy.* Annapolis, MA: Navy Institute Press.

Pandey, M. D. and Nathwani, J. S. 2003. A conceptual approach to the estimation of societal willingness-to-pay for nuclear safety programs. *International Journal of Nuclear Engineering and Design,* 224, 65–77.

Perrin, C. 2005. *Shouldering Risks: The Culture of Control in the Nuclear Power Industry*. Princeton, NJ: Princeton University Press.

Perrow, C. 1999. *Normal Accidents: Living with High Risk Technology*. Princeton, NJ: Princeton University Press.

Rasch, G. 1980. *Probabilistic Models for Some Intelligence and Attainment Test*. Chicago, IL: University of Chicago Press.

Rasmussen, J. 1979. *On the Structure of Knowledge—A Morphology of Mental Models in a Man-Machine Context*. Roskilde, Denmark: RISOM-2192, RISO National Laboratory.

Rasmussen, J. 1997. Risk management in a dynamic society: A modelling problem. *Safety Science*, 27(2/3), 183–213.

Read, R. 2012. How Tenacity, a Wall Saved a Japanese Nuclear Plant from Meltdown after Tsunami. Available at: http://www.oregonlive.com/opinion/index.ssf/2012/08/how_tenacity_a_wall_saved_a_ja.html.

Reckard, E. S. 2011. Fukushima nuclear plant owner is slammed for lacking of candor. *Los Angeles Times*, March 21, 2011.

Rees, J. V. 1994. *Hostages of Each Other*. Chicago, IL and London, UK: University of Chicago Press.

Rickover, H. 1983. Review of the General Public Utilities Nuclear Corporation organization and senior management competence after TMI #2 accident to operate TMI Unit #1. Report by Admiral H. G. Rickover, USN, November 19, 1983 for GPUN.

Rogovin, G. T. 1980. *Three Mile Island, Volume II, Parts 1, 2, and 3*. A Report to the Commissioners and the Public. Washington, DC: Nuclear Regulatory Commission.

Saeed, K. 2015. Jay Forrester's operational approach to economics. *Systems Dynamics Review*, 30(4), 233–261.

Sandy. 2012. "Hurricane Sandy" at the Northeast Coast October 29th, 2012. Available at: www.weather.gov.

Spurgin, A. J. 1994. Developments in the use of simulators for human reliability and human factors purposes. *IAEA Technical Meeting on Reliability Analysis and Probabilistic Safety Assessment*, Budapest, Hungary.

Spurgin, A. J. 2009. *Human Reliability Assessment: Theory and Practice*. Boca Raton, FL, London, UK, and New York, NY: CRC Press, Taylor & Francis Group.

Spurgin, A. J. 2013a. *Human Reliability Assessment: Theory and Practice* (In Japanese; Tomoaki Uchiyama, trans.). Japan: SIB Access, Co. Ltd.

Spurgin, A. J. 2013b. Application of Cybernetic Models in the Study of Safety and Economics of Nuclear Power Systems and Other High Risk Organizations. PhD thesis, City University, London, UK.

Spurgin, A. J. and Carstairs, R. L. 1967. Overall station control at Hunterstone A. *Proceedings of the Electrical Engineers*, 114(5), 671–678.

Stolberg, S. G. et al. 2015. Amtrak train derailed going 160 MPH on sharp curve; at least 7 killed. *New York Times*, May 13, 2015.

Straeter, O. 2000. *Evaluation of Human Reliability on the Basis of Operational Experience*. Kohl, Germany: GRS-170, GRS.

Sursock, J. P. and Lewis, S. 2015. *An approach to risk aggregation for risk-informed decision-making*. Report # 3002003116, April. Palo Alto, CA: Electric Power Research Institute.

Swain, A. D. and Guttman, H. E. 1983. *Handbook of human reliability analysis with emphasis on nuclear power plant applications*. NUREG/CR-1273. Washington, DC: US Nuclear Regulatory Commission.

Thames Barrier. 1982. Available at: Wikipedia.org/wiki/Thames_Barrier.

Thomas, P. 2013. Methods for measuring risk-aversion: Problems and solutions, Joint IMEKO TC1-TC7-TC13 Symposium, J. Phys.: Conf. Ser., 459, 012019.

Thomas, P. 2014. The J-value framework for determining best use of resources to protect humans and the environment. Invited lecture at the *First International Conference on Structural Integrity (ICONS-2014),* February 4–7, 2014, Kalpakkam, India.

Thomas, P. and Jones, R. 2010. Extending the J-value framework for safety analysis to include the environmental costs of a large accident. *Process Safety and Environmental Protection*, 88, 297–317.

Thomas, P., Jones, R., and Boyle, W. 2010. The limits to risk aversion: Part 2. *Process Safety and Environmental Protection*, 88, 396–406.

Thomas, P. and Stupples, D. 2006. J-value: A universal scale for health and safety spending. Special feature on systems and risk. *Measurement + Control*, 39/9, 273–276.

Thomas, P. J., Kearns, J. O., and Jones, R. D. 2010. The trade-offs embodied in J-value safety analysis. *Process Safety and Environmental Protection*, 88(3), 147–167.

Thomas, P. J., Stupples, D. W., and Alghaffar, M. A. 2006a. The extent of regulatory consensus on health and safety expenditure. Part 1: Development of the J-value technique and evaluation of the regulators' recommendations. *Process Safety and Environmental Protection*, 84(5), 329–336.

Thomas, P. J., Stupples, D. W., and Alghaffar, M. A. 2006b. The extent of regulatory consensus on health and safety expenditure. Part 2: Applying the J-value technique to case studies across industries. *Process Safety and Environmental Protection*, 84(5), 337–343.

Thomas, P. J., Stupples, D. W., and Alghaffar, M. A. 2006c. The life extension achieved by eliminating a prolonged radiation exposure. *Process Safety and Environmental Protection*, 84(5), 344–354.

Trucco, P., Leva, M., and Straeter, O. 2006. *Human Prediction in ATM via Cognitive Simulation: Preliminary Study*. New Orleans, LA: PSAM 8.

Walker, J. 1991. *The Viable Systems Model: A Guide for Cooperatives and Federations. Manual.* Part of a Training Package for Strategic Management for Social Economy (SMSE) carried out by ICOM, CRI, CAG and Jon Walker.

WASH 1400. 1975. Reactor safety study: An assessment of accident risks in US commercial nuclear power plants. NUREG-74/014. Washington, DC: US Regulatory Commission.

# Index